ZRAC

JUN -- 2020

SCIENTISTS
WHO CHANGED HISTORY

SCIENTISTS

WHO CHANGED HISTORY

DK LONDON

SENIOR EDITORS: Victoria Heyworth-Dunne,
Kathryn Hennessy
SENIOR ART EDITORS: Stephen Bere, Mark Cavanagh
LEAD ILLUSTRATOR: Phil Gamble
EDITORS: Rose Blackett-Ord, Kim Bryan,
Andy Szudek, Debra Wolter
EDITORIAL ASSISTANT: Daniel Byrne
US EDITOR: Kayla Dugger
PICTURE RESEARCH: Sumedha Chopra
JACKET DESIGNERS: Priyanka Bansal, Surabhi
Wadhwa-Gandhi
JACKET EDITOR: Emma Dawson
JACKET DESIGN DEVELOPMENT MANAGER: Sophia MTT
SENIOR DTP DESIGNER: Harish Aggarwal
PRODUCER, PRE-PRODUCTION: Andy Hilliard
PRODUCER: Rachel Ng
MANAGING EDITOR: Gareth Jones
SENIOR MANAGING ART EDITOR: Lee Griffiths
ASSOCIATE PUBLISHING DIRECTOR: Liz Wheeler
ART DIRECTOR: Karen Self
DESIGN DIRECTOR: Philip Ormerod
PUBLISHING DIRECTOR: Jonathan Metcalf

CONSULTANT: Chris Woodford
CONTRIBUTORS: Alexandra Black, Alethea Doran,
Joanna Edwards, Richard Gilbert, Janet Mohun,
Victoria Pyke, Penny Warren

First American Edition, 2019
Published in the United States by DK Publishing
1450 Broadway, Suite 801, New York, NY 10018

Copyright © 2019 Dorling Kindersley Limited
DK, a Division of Penguin Random House LLC
19 20 21 22 23 10 9 8 7 6 5 4 3 2 1
001–312740–Sept/2019

A catalog record for this book
is available from the Library of Congress.

ISBN 978-1-4654-8248-8
Printed and bound in China

A WORLD OF IDEAS:
SEE ALL THERE IS TO KNOW

www.dk.com

CONTENTS

CONSULTANT: **Chris Woodford**
Chris Woodford graduated from Cambridge University with a degree in Natural
Sciences and has written and edited many books with Dorling Kindersley,
including the best-selling *Cool Stuff* series. His most recent book, *Atoms Under the
Floorboards*, won the 2016 American Institute of Physics Science Writing award.

4

SCIENCE AND INDUSTRY

1800–1895

5

PARADIGM SHIFTS

1895–1925

6

WAR AND MODERNITY

1925–1950

7

THEORIES OF EVERYTHING

1950–present

INTRODUCTION

Science is humanity's ongoing attempt to understand how the Universe works. Whether in the form of meticulous research, ingenious insight, or unexpected discoveries, the search for truth continues to challenge, often leading to more questions than answers.

Throughout history, humans have always been driven to understand the workings of the world. In their quest for knowledge, philosophers in ancient Greece were the first to try to explain what they observed. However, many of their ideas were inaccurate, as their philosophical method lacked any experimental evidence. Thales of Miletus in the 6th century BCE, for example, proposed that water is the primary substance of the cosmos, as he had realized that water is essential for life and that it exists on land, sea, and in the air. Two centuries later, Aristotle wrote widely on scientific subjects, from physics and biology to astronomy, laying the foundations for much of the work that has followed. However, he also made fundamental errors, such as arguing that heavier objects fall faster than lighter ones, because he relied on thought and argument rather than empirical proof.

Despite being some distance from a reliable scientific method—in which a hypothesis is proposed, systematically tested, and evaluated on the strength of the results—the early thinkers made vital findings. Halley's comet was observed in China in 240 BCE, and some 300 years later, Zhang Heng explained eclipses and drew up an extensive catalog of stars.

Scientific dawn

While European progress in science stalled during the Middle Ages, the shift of knowledge to the Islamic world inspired rapid advances in scientific thinking. The move of most of the important writings from Greece and India to The House of Wisdom—the library of the Abbasid caliphate in Baghdad—in the 8th century CE nurtured scholars. The scholars included the mathematician al-Khwarizmi, whose works included trigonometry, algebra, and astronomy, and Alhazen, an innovator in the field of optics, whose studies of dissected bulls' eyes were the first scientific experiments.

Progression of ideas

A mark of a true scientist is the ability to evaluate and revise previously held truths. Today, scientists understand that science is never finished and that their work might be superseded. In 1590, a major revision occurred when Galileo Galilei disproved Aristotle's theory of falling bodies, instead asserting that objects fall at the same finite speed regardless of mass. Galileo's support for Copernicus's radical idea of a heliocentric universe, where the Earth rotates around the Sun, led to Galileo being declared a heretic, but it also paved the way for the modern understanding of the Universe.

Science comes of age

During the 17th century, Isaac Newton precipitated the scientific revolution of the early modern period with his laws of motion and law of universal gravitation, which fundamentally changed the way that humans understood the world. Progress gathered pace during the 18th century in diverse fields, such as geology, with James Hutton's theory of the ancient age of the Earth, and chemistry, with John Dalton's discovery of the nature of atoms. Alessandro Volta's electric battery immediately benefited society and provided a new tool for scientists. Michael Faraday's study of electromagnetism led to the invention of the electric motor, a utilitarian and epoch-making device.

In search of truth

Science increasingly changed the course of everyday life through the 20th century. Alexander Fleming's discovery of penicillin in 1928 saved at least 200 million lives

"The important thing is **not to stop questioning.**"

Albert Einstein, 1955

during World War II, while Marie Curie's work in the field of radioactivity has had a lasting impact on medical and scientific research. Meanwhile, Albert Einstein's theories of relativity rewrote the rules of classical physics and introduced entirely new concepts to the study of the nature and origins of the Universe.

Advances, such as even more powerful computers, as conceived by Alan Turing, and better connected ones, realized by Tim Berners-Lee's World Wide Web, have enabled scientists to move faster and further in the pursuit of knowledge. Frederick Sanger unraveled the chemical sequence of DNA, while Stephen Hawking used relativity and quantum theory to predict the physics of black holes.

From the makeup of our chromosomes to the deepest reaches of space-time, scientists will always strive for truth. Building on previous work and theories, they will continue to shape history and seek to answer the most essential questions about the Universe.

1

THE DAWN OF SCIENCE

650 BCE–1450

THALES

Astronomer and mathematician Thales of Miletus is the first person in Europe known to have turned away from mythology and used observation to understand the natural world. He effectively invented empirical science and has since become known as the founder of Western philosophy.

MILESTONES

RADICAL THINKER
Establishes a new scientific way to explain phenomena; founds the Milesian School of philosophy.

BASIC SUBSTANCE
Concludes that everything in the Universe derives from a single substance, namely water.

FLOATING EARTH
Proposes that the Earth is a disk surrounded by water, which solidifies to form the land.

APPLIED GEOMETRY
Uses geometrical principles to calculate the height of one of the Egyptian pyramids.

Thales was born in the affluent city-state of Miletus on the coast of modern-day Turkey during the 7th century BCE. Little is known about his life, and what is known derives from ancient Greek scholars writing centuries later, such as Aristotle (see pp.14–17); the historian Herodotus; and the 3rd-century biographer of the Greek philosophers, Diogenes Laërtius. There are no surviving records of Thales' own work or writings. However, according to available source material, Thales was the first of a new generation of thinkers who established a rational, scientific approach to explaining the natural world.

Forging new paths
During Thales' lifetime, superstition and a fear of the gods dominated Western civilization—catastrophic events were ascribed to the wrath of deities who had to be appeased. Yet in an unprecedented break with convention, Thales sought rational explanations rather than religious ones for natural phenomena. He believed that every event,

Thales was one of the Seven Wise Men of antiquity—a group of intellectuals and orators from the 7th and 6th centuries BCE, esteemed for their wisdom.

"Everything is made of water."

Thales of Miletus, 7th–6th century BCE

such as a crop failure or a flooding, resulted from natural causes that could be identified through observation and reason. Further, he believed that if events could be rationally explained and understood, they could also be predicted. This approach to life marked a significant shift in human thinking.

The nature of the Universe

Another fundamental leap made by Thales was to question the nature of the Universe and what constitutes matter. He believed that there must be a single substance from which everything in the cosmos derives. Thales reasoned that water was the most likely candidate because it was vital to all forms of life, was capable of motion, and existed in different forms (solid, liquid, gas). He concluded that the substance in question must be water.

He also questioned the position of the Earth in the Universe and made the cosmological proposition that it must exist as a flat disk floating on water. His theory rested on the logic that every land mass is surrounded by water (the sea) and that animals, plants, and the Earth itself all absorb water. The significance of both theorems lies not in Thales' conclusions, which were incorrect, but in his desire to question the world and explain it rationally.

Applying science to life

Thales is praised for his grasp of mathematics and geometry. In an example cited by Herodotus, he calculated the height of an Egyptian

> ## "**Day** became **night**, and this change... **Thales** the Milesian had **foretold.**"
>
> **Herodotus**, 5th century BCE

pyramid by measuring the length of his own shadow and that of the pyramid at the same time of day, then applying the ratio he found between his shadow length and his height to that of the pyramid's shadow.

REASON

Thales also used geometric principles to determine the distance of a ship from the shore. One of his most impressive yet controversial feats, as alleged by Herodotus, was to have correctly predicted a total solar eclipse, which took place on March 28, 585 BCE. This prediction would have required Thales to identify an 18-year cycle in the movements of the Sun and the Moon (the Saros cycle), which many believe casts doubt on Herodotus's claim.

Legacy

The first person in Western society recorded as having engaged in rational and scientific thought, Thales inspired others to follow his approach through his questioning mind and his application of deductive reasoning. He initiated a new wave of thinking; his teachings sparked debate and counter-theorems, and form the origin of modern scientific thought.

Applying observation and reason to questions about the natural world, Thales overturned centuries of traditional views and laid the foundations of empirical science.

PYTHAGORAS

The renowned ancient Greek Pythagoras was a mathematician and philosopher.

Inspired by Thales, Pythagoras (c.570–500 BCE) believed that the world is mathematical in nature. Aged 30, he set up a commune that pursued mathematical research and mysticism. While his work is known only through his disciples, he made key contributions to Western thought, including the idea that the Earth and the planets orbit around a central "Hearth." He had a profound influence on both Plato and Aristotle.

CREDITED WITH **DISCOVERING 5** GEOMETRIC THEOREMS

CORRECTLY PREDICTED AN **ECLIPSE,** ENDING A **15-** YEAR WAR

CONSIDERED **1** OF THE **7 WISE MEN** OF ANCIENT **GREECE**

OBSERVATION

Known as both a philosopher and a scientist, Aristotle laid the foundations for scientific methodology by introducing the concept of logic. Using reasoning to support explanations about natural phenomena, his writings spanned almost every field of science. His ideas held sway for centuries and were profoundly influential on both Western and Arabic thinking.

MILESTONES

PLATO'S PUPIL
Joins Plato's Academy in 367 BCE, and remains there until his mentor's death 20 years later.

BIOLOGICAL STUDIES
In 348 BCE, begins to write *On the Parts of Animals* and *The Generation of Animals*, naming 500 species.

ROYAL CHARGE
Becomes tutor to 13-year-old Prince Alexander of Macedon, later Alexander the Great, in 342 BCE.

PROLIFIC WORKS
Founds the Lyceum in 335 BCE, and begins researching and writing many of his treatises.

Originally from Macedonia, northeastern Greece, Aristotle traveled to Athens at the age of 17 to study at the Academy of Plato, the center of learning in the Greek world. When Aristotle left, he used what he had learned there to develop an approach that was based more on observation than on theory. For Aristotle, facts about the natural world were as interesting as the process of reasoning; in this way, he was both one of the first scientists and a philosopher.

Aristotle realized that rational thinking was fundamental to science, and he developed an ordered approach to logical reasoning. Moreover, he understood that science requires practical data, which is gained by observation. Aristotle saw that such information can form the premises (starting points) from which conclusions can be derived about the world by means of logic.

Aristotle's work was extraordinarily wide-ranging in its subject matter. He wrote some 200 treatises, which developed the framework for the fields of physics, chemistry, biology, astronomy, logic, and mathematics. One of his most important surviving written works, *Physics*, remained unchallenged until the Middle Ages. He believed that the science of physics encompassed almost everything there is

Aristotle's Lyceum was known as the "Peripatetic School," after the Greek peripatoi, *meaning "walking," because the teaching was conducted while strolling around.*

" For nature is everywhere **the cause of order.**"

Aristotle

ARISTOTLE

384–322 BCE

BELIEVED THERE WERE 5 ELEMENTS, INCLUDING 1 "HEAVENLY ELEMENT," QUINTESSENCE

IDENTIFIED 500 LIVING SPECIES AND 11 GENERA OF ANIMALS

31 OF ARISTOTLE'S 200 WORKS SURVIVE

THEOPHRASTUS

A student and close colleague of Aristotle's, Theophrastus was his successor at the Lyceum.

Like Aristotle, Theophrastus (c.371–287 BCE) had wide-ranging knowledge, but his special interest was botany. He described more than 500 species of plant, identified the role of leaves in plant nutrition, and was the first to use botanical nomenclature. When Alexander the Great sent back plant samples from his invasions, Theophrastus documented them and planted cuttings, thereby introducing species such as rice and cotton to Greece.

to know about the world. Aristotle clearly regarded physics—and other science subjects, such as zoology—just as highly as he did philosophical subjects such as ethics and metaphysics. However, at that time, there was still a certain amount of overlap between the subjects, and it was not until the 17th century that philosophy and science were really considered to be distinct.

The father of many sciences

Applying his principle that all findings should be based on observation and reasoning, Aristotle made significant discoveries in a variety of subjects. In astronomy, for example, he disproved the long-held belief that the Earth was flat. Having observed several

Aristotle's legacy endures to this day. His methodical and analytical approach to the pursuit of truth was the starting point for science in its current form, and he believed in learning for its own sake.

lunar eclipses, he noted that the Earth's shadow on the Moon was always curved. This led him to reason, "since it is the interposition of the Earth that makes the eclipse, the form of this line will be caused by the form of the Earth's surface, which is therefore spherical."

In the subject of physics, Aristotle laid the groundwork for later advances by Galileo (see pp.54–55) and Newton (see pp.76–81). Although Galileo disproved many of Aristotle's theories (for example, that the Moon and celestial bodies were perfectly smooth and unchanging), Aristotle's writings were the starting point for Galileo's investigations into physics and astronomy. Galileo adapted Aristotle's principle that force is required

for steady motion to establish his own theory: that force changes motion but is not essential to sustain it (the concept of inertia). Newton, in turn, built on this and formalized it into his first law of motion. He also adopted Aristotle's view of time as uniform and ever-flowing.

Aristotle also pioneered the scientific study of plants and animals. He was fascinated by the idea of classifying living things and used empirical investigation to develop theories about the animal world. In *The Generation of Animals*, for example, he proposed that animals develop in their mother's womb or in eggs. Around one-third of Aristotle's surviving texts are on zoology.

Like his mentor Plato, Aristotle was also committed to education. He set up the Lyceum in Athens, which became the first institute to offer a comprehensive education in all areas of knowledge, from physics to politics, drama, and poetry: similar in scope to a modern university.

"Our observations of the stars make it evident ... that the Earth is circular."

Aristotle

ARCHIMEDES

An exceptional theoretician, Archimedes was one of the greatest mathematicians of the ancient world. Applying his genius to solve practical problems, his original ideas gave rise to inventions, as well as writings, in geometry, mathematics, mechanics, and physics that are still valuable today.

MILESTONES

STUDENT WORKS
In Alexandria, c.250 BCE, writes the works *On the Measurement of a Circle* and *On Floating Bodies*.

BUILDS SHIP
In c.240 BCE, designs *Syracusia*, a giant ship built as a gift for Ptolemy III of Egypt. It sails once.

NUMBER SYSTEM
Devises a way of writing big numbers and calculates the number of sand grains in the Universe.

EUREKA MOMENT
Taking a bath, he finds how to measure the volume of irregular objects, and "Archimedes' principle."

Born in Syracuse, a Greek city-state in modern-day Sicily, Archimedes followed his father Phidias by studying astronomy and mathematics. Early in his career, he went to Egypt to study in Alexandria, the center of learning in the ancient world. Here, he may have known such brilliant minds as Eratosthenes of Cyrene (see p.21), to whom he addressed his *Method of Mechanical Theorems*. Archimedes then returned to Syracuse, where King Hieron II—who may have been a relative—became his patron.

Geometry genius
Archimedes was a polymath who turned his intellect to a host of problems, but his pet subject was geometry. Such was his devotion to the study of shapes that he saw his inventions as trivial in comparison. He was the first to calculate the approximate value of π (pi)—the ratio of a circle's circumference to its diameter—and in *On Conoids and Spheroids*, he set out how to calculate the volume of geometrical solids. Archimedes considered one of these mathematical proofs—that the volume of a

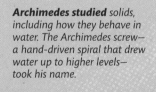

Archimedes studied solids, *including how they behave in water. The Archimedes screw— a hand-driven spiral that drew water up to higher levels— took his name.*

sphere encased by a cylinder is two-thirds that of the cylinder—to be his greatest achievement, and he asked for the shapes to be carved on his tombstone.

Royal assignment

The Roman writer Vitruvius recorded a possibly apocryphal story of how one of Archimedes' most important discoveries was made. King Hieron II suspected that his new crown was not pure gold, so he asked his chief scientist to investigate. Having weighed the crown, Archimedes realized that if he also knew the crown's volume, he could work out its density—and since metals have different densities, he could prove whether it was pure gold.

Archimedes pondered how to measure the volume of irregular objects on a visit to the public bath house. As he got into the tub, he noticed the water rising and spilling over the side, and he felt lighter. He is said to have shouted *"Eureka!"* ("I found it!") and run home naked. He had realized that the volume of water spilled from the bath was equal to his body's volume, and that an object immersed in a liquid receives an upward force equal to the weight of the liquid it displaces.

In practice, the effect of water spilled by the crown would have been small and difficult to measure. Instead, Archimedes may have measured the upthrust by hanging the crown and an equal weight of gold on both ends of a horizontal stick, which he then lowered into water. If the crown was pure gold, both objects would receive an equal upthrust and the stick would stay level.

Military mechanics

"Archimedes' principle," as it came to be known, was also put to militaristic use. Archimedes used it to design

Syracusia, a 2,000-man ship given as a gift to Ptolemy III, and devised a system of levers and pulleys for launching the vessel. The royal fleet also benefited from the Archimedes screw, a device that he may have seen in Egypt for pumping water up from the bilges of ships. When Syracuse was besieged by the Romans in 213 BCE, Archimedes invented the Claw of Archimedes—a mechanical hook designed to lift an enemy ship by the prow, throwing sailors into the sea and capsizing or damaging the vessel. Despite orders that he should not be harmed, he was killed by a Roman soldier in 212 BCE.

> "Give me a **lever** and a **place to stand** ... **and I'll move the Earth."**

Archimedes

Archimedes' findings on measuring volume and the buoyancy of objects in a fluid proved that Hieron II's crown was not pure gold, but an alloy of gold and other, less dense metals.

ERATOSTHENES OF CYRENE

Born in Cyrene in North Africa, Eratosthenes became chief librarian at the library of Alexandria in Egypt, the greatest center of learning in the ancient world, at the age of 40.

Eratosthenes (276–194 BCE) drew maps of the world, calculated the distance from the Earth to the Sun, devised a reliable method for finding prime numbers, and coined the term "geography." He was most renowned for accurately calculating the Earth's circumference. He measured the angle of a shadow at Alexandria as 7.2°, while 500 miles (800 km) away at Swenet, the Sun was directly overhead. As 7.2° is one-fiftieth of the circumference of a circle (360°), he reasoned that the two points must be one-fiftieth of Earth's circumference apart. He multiplied that distance by 50 to get 25,000 miles (40,000 km)—an error of less than 2 percent.

PROVED
A SPHERE'S VOLUME IS
TWO-THIRDS
THAT OF A CYLINDER ENCASING
IT EXACTLY

SYRACUSIA WEIGHED
4,000 TONS

FOUND VALUE OF π (PI) TO BE
C.3.142

An astronomer, inventor, and mathematician, as well as a renowned writer and artist, Zhang Heng was one of ancient China's greatest thinkers. Among his scientific innovations was the first device for detecting earthquakes.

Born during the Eastern Han Dynasty, whose rulers promoted science and technology, Zhang Heng achieved success as a literary scholar before turning his attention to astronomy and mathematics. The expertise he acquired in those fields led to appointments at the court of Emperor Shun, where he rose to become chief astronomer.

Discoveries and inventions

From his thorough observations of the night sky, Zhang concluded that the Moon and planets did not produce light, but reflected it from the Sun; he also believed that the Universe and planets were spherical. He mapped the stars and planets in a detailed catalog, identifying 124 constellations; 2,500 large stars; and 11,250 smaller ones.

A highly skilled engineer, Zhang constructed a water-powered armillary sphere: a model that depicted the night sky through a framework of rings, positioned to mirror the movements of celestial objects. He also designed a nonmagnetic compass and developed an instrument that recorded the direction in which an earthquake had occurred—this is regarded as the world's first seismoscope.

The basic principles of Zhang's earthquake-registering instrument subsequently featured in studies that led to the development of modern seismographs. His most notable contribution to the field of mathematics was his own calculation of the formula for π (pi); his figure would not be improved upon in China for another 200 years.

Zhang's seismoscope traced Earth tremors. On detection, a ball made a sound by dropping from one of the eight dragons' heads into the frog below it.

"The **sky** is like a hen's egg ... the **Earth** is like the **yolk** of the egg."

Zhang Heng, 120 CE

78–139 CE

ZHANG HENG

GALEN

The greatest medical practitioner of his era, Claudius Galenus, better known as Galen, was physician to three Roman emperors, a skilled surgeon and anatomist, and a prolific author. His work took Greco-Roman medical knowledge to its highest level, and his ideas prevailed until the Renaissance, 1,300 years after his death.

MILESTONES

FIGHTERS' DOCTOR
Appointed physician to the gladiators by the high priest of the temple of Pergamon in 159 CE.

PLAGUE IN ROME
Treats the victims of the Antonine plague, thought to be smallpox, which breaks out in 166 CE.

LEAVES THE CAPITAL
Stays away from Rome from 166 CE to 169 CE due to his unpopularity with the established physicians.

STATE APPOINTMENT
Spends four decades as physician to emperors, beginning with Marcus Aurelius in 169 CE.

PROLIFIC WRITER
Writes more than 500 works, dictating to scribes. Around 100 survive a fire in his storerooms in 191 CE.

Galen was born in the eastern part of the Roman Empire, in Pergamon on the Aegean coast, which is now the Turkish city of Bergama. His father, Aelius Nicon, was a wealthy Greek architect who wanted his son to pursue a career in politics. However, it is said that when Galen was 16, the Greek god of healing, Asclepius, came to his father in a dream requesting that Galen study medicine, and he duly began his medical training at the healing temple (Asclepeion) in Pergamon.

Three years later, Galen's father died, and he used his inheritance to travel. He spent 5 years studying in Alexandria, Egypt, at the most advanced medical school in the Roman Empire. On his return to Pergamon, he worked as physician to the Roman gladiators. Galen wrote that the gladiators' wounds were as "windows into the body." Since 150 CE, Roman law had forbidden the dissection of human corpses, so these injuries gave him valuable practical insights into anatomy.

Imperial physician

In 162 CE, Galen went to the city of Rome, where he gave lectures and demonstrated his knowledge and surgical dexterity by dissecting pigs and apes. He also treated a number of eminent statesmen and caught

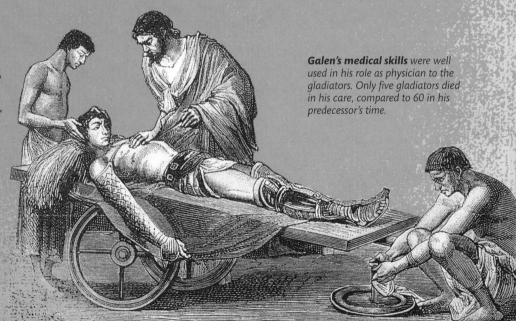

Galen's medical skills were well used in his role as physician to the gladiators. Only five gladiators died in his care, compared to 60 in his predecessor's time.

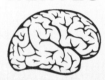

the attention of the emperor, Marcus Aurelius. In 169 CE, Aurelius asked Galen to be his personal physician, describing him as "first among doctors and unique among philosophers." Galen was later the physician of the next two emperors: Lucius Commodus and Septimius Severus.

Medical theory and practice

In his book *On Anatomical Procedures*, Galen spoke of the importance of dissection and observation. Although his experience of dissection was limited to animals, he made some very important discoveries, including findings relating to the functions of many organs. The great cerebral vein (in the skull), which he identified, is still known as "the vein of Galen."

Galen believed that the body had three connecting systems: the brain and nerves; the heart and arteries, which he said were responsible for "vital energy"; and the liver and veins, where he said blood was formed and distributed around the body. He dispelled some misconceptions, revealing that the kidneys, not the bladder, make urine; that the heart has valves; and that arteries carry blood, not air, as was previously thought. He worked out that the spinal cord transmits messages to the brain and identified which nerves are involved in movement and speech.

Treating many illnesses such as malaria (which he called "fever") and smallpox (known as "plague"), Galen set great store by listening to patients and observing them, including taking a detailed record of their pulse—a procedure he described in his book *De Pulsibus*. He believed that diet, exercise, and good hygiene were important, and thought there was a strong connection between the

mind and the body, saying that anger could exacerbate fever and anxiety could result in insomnia.

The ancient Greeks believed that health was influenced by the four "humors": bodily fluids that caused a person to be sanguine, choleric, phlegmatic, or melancholic. While Galen accepted this idea, he also expanded on it. Unlike

Hippocrates (see box), he believed the "humoral imbalance" could be located in a single organ. To treat such problems, he used herbs. He cataloged more than 300 plants with medicinal properties and developed a system known as the "Galenic degrees" to describe their potency. He also practiced bloodletting (or "breathing a vein"), believing that fever and headaches came from too much "sanguine humor."

Galen is thought to have lived to the age of 87. His ideas formed the cornerstone of medical thinking for centuries.

> **"The working body is not understandable without knowledge of its structure."**
>
> **Galen**

It was illegal to dissect human bodies in Rome, so Galen had to learn what he could from dissecting apes. This caused him to make a few mistakes about human anatomy and the workings of the body.

HIPPOCRATES

Known as the "Great Physician," Hippocrates was the first physician to develop a rational system for the treatment of disease.

Unlike his contemporaries, Hippocrates (c.460–370 BCE) did not believe that diseases had supernatural causes. He thought that ill health resulted from imbalances in the body's four humors and that the physician's role was to restore that balance. He founded a medical school on his native island of Kos, where new students swore an oath to maintain high ethical standards, including the principles of confidentiality (to keep a patient's illness private) and nonmaleficence (to ensure that a patient is never harmed). Swearing the "Hippocratic oath" is still common practice in many countries.

DE QVO PIVS ET INEST LOM̄ LEXA ODIERENIVS

GALIENVS   O IPEPAS

MVNDIPRE
SENTIS SE
ＥＭＡＮＴ
ＺＥＬＥＶＮＴＳ

EXHIS FOR
MANTROVE
NIQVI
EVOETREAL VE

...ATV...SMAGNIS P̄M L I DOL M̄...IS

"AS POLES TO TENTS AND WALLS TO HOUSES, SO ARE BONES TO LIVING CREATURES … I WOULD HAVE YOU GAIN AN EXACT AND PRACTICAL KNOWLEDGE OF HUMAN BONES."

Galen
On Anatomical Procedures

◄ **This fresco of Galen and Hippocrates** *dates from the mid 13th century.*

The earliest recorded female mathematician, Hypatia advanced the concept of conics, which underpins geometry and astronomy, and has important applications in everything from bridge building and architecture to satellite navigation.

MILESTONES

SCHOOL DIRECTOR
Becomes director of Alexandria's Neoplatonic School in c.400 CE, where she also teaches.

SEMINAL WORKS
Writes commentaries on Apollonius's *Conics* and Diophantus's *Arithmetica* in the early 400s CE.

SCIENTIFIC MARTYR
Perceived as a symbol of scientific paganism; dies at the hands of a conservative Christian mob in 415 CE.

TEXTS PRESERVED
Work survives to the 10th century and is recorded in the *Suda*, an encyclopedia of the Byzantine world.

ARTISTIC TRIBUTE
She is the only woman depicted in Raphael's 1511 Renaissance painting *School of Athens*.

Hypatia's mathematical genius may in part be explained by the outstanding education she is thought to have received from her father, Theon, a professor of mathematics at the University of Alexandria in Egypt, then part of the Roman Empire. She collaborated with her father on a treatise about Greek mathematician Euclid and probably helped him revise a groundbreaking 13-volume work, Euclid's *Elements*, which sets out extensive constructions and proofs for geometry, proportion, and number theory. Hypatia may also have assisted her father in preparing his edition of *The Almagest* by the ancient Greek mathematician Ptolemy. As well as teaching his daughter mathematics, astronomy, astrology, and philosophy, Theon also impressed upon her the significance of teaching, and she developed into a gifted speaker. This was an important factor in one of Hypatia's major achievements: the popularization of mathematics.

Math and mechanics
Although she also lectured on astronomy, philosophy, and mechanics, Hypatia became best known for her mathematics work—in particular, her explanation of the conic sections introduced by Apollonius in 200 BCE. Conic sections are one of the most important functions of geometry; they are produced when a plane intersects a cone. Depending on the angle, the resulting conic section may be a circle, ellipse, parabola, or hyperbola. Hypatia edited *On the Conics of Apollonius*, which elaborated on conic sections and made them easier to understand and apply. As a result, conics became firmly entrenched in the mathematical repertoire.

Hypatia may have developed the astrolabe, an astronomical calculating device used to measure the distances of stars to establish the user's latitude.

HYPATIA

Hypatia's work was later expanded on by eminent scientists, including Isaac Newton, who used conic sections to realize his theories of planetary movement (see pp.76–81).

Skilled in algebra, Hypatia wrote a commentary on the *Arithmetica of Diophantus*, helping to explain the principles of solving algebraic equations. Parts of her commentary are thought to have been reworked into a later Arab edition of *Diophantus*, which survived the early Middle Ages and reemerged in the Renaissance, influencing Pierre Fermat's development of modern number theory.

In addition to her mathematics work, Hypatia has been associated with the development of several mechanical devices. In the preserved letters of Synesius of Cyrene—later Bishop of Ptolemais, one of her most devoted students—are instructions for constructing an astrolabe, a navigational instrument whose design was advanced enough to measure the positions of the planets, the stars, and the Sun. Synesius claimed that Hypatia invented the astrolabe, although it is more likely that she made improvements to the existing design (see box). Another device attributed to Hypatia in the letters is a hydrometer for measuring the specific gravity of a liquid.

Enduring legacy

Regarded as the most celebrated female scientist of the ancient world, Hypatia's advances were significant not only because of her gender, but also because she was one of the last scientists of note before the early Middle Ages, when scientific endeavors in Europe were largely stamped out by Christian zealots who believed math and science to be heresy. Hypatia lived during the final years of the Roman Empire, and her violent death is regarded as a result of growing Christian fervor against rational thought and those, like Hypatia, who were pagans. While scholarship might have gone into decline after Hypatia's death, centuries later, her ideas were to inform the work of René Descartes, Isaac Newton, and Gottfried Leibniz, among others.

> **"**Hypatia ... made such attainments in literature and science, as to **far surpass all the philosophers of her own time."**
>
> **Socrates**, c.450 CE

An exceptional educator, Hypatia was adored by her students. She lectured widely and popularized interest in mathematics and astronomy.

PTOLEMY

The most influential astronomer and geographer of his time, Ptolemy outlined a model of the Universe that reigned for more than 1,400 years.

Among Ptolemy's (c.100–c.170 CE) key works was *The Almagest*, which explained the motions of the heavenly bodies mathematically, and *Geographica*, an atlas containing instructions for map making. He devised several armillary spheres for measuring the angles of the stars, the planets, and the Sun, notably the astrolabe. Centuries later, Islamic astronomers used the astrolabe, and later still European astronomers. Although his model of the Universe put the Earth at its center, his mathematical approach was the starting point for future generations of astronomers.

HELPED
TO REVISE THE
13 BOOKS
OF EUCLID'S
ELEMENTS

HER COMMENTARY
ON **DIOPHANTUS'S**
ARITHMETICA
CONTAINED
MORE THAN
100
MATHEMATICAL
PROBLEMS

In the 9th century, the Muslim scholar al-Khwarizmi developed the Hindu-Arabic number system and made important advances in algebra. These innovations underpinned the subsequent development of trade, technology, and science.

MILESTONES

EXCELS IN ACADEMIA
Becomes director of the House of Wisdom under Caliph Al-Ma'mun in the early 800s CE.

CARTOGRAPHIC WORK
Helps compile a world map; leads a project to measure the circumference of Earth between c.811 and 825 CE.

TRANSFORMS MATH
Writes a book on numerals in c.825 CE; its Latinized title is the origin of the word "algorithm."

ADVANCES ALGEBRA
Publishes the world's first algebra book in c.830 CE; it includes the theory of quadratic equations.

During his life's work at the House of Wisdom, a center of scientific research and teaching in Baghdad (now in Iraq), Muhammad ibn Musa al-Khwarizmi wrote influential treatises on cartography and astronomy, along with two books that had profound consequences for Arabic and Western mathematics. Adapting an existing Indian numbering practice, he proposed the figures 0 and 1 to 9 as the basis of a new number system and showed how to do arithmetic with them. This system, which became known as Hindu-Arabic numerals, was introduced to Europe in the 12th century through a Latin translation of the book *Al-Khwarizmi Concerning the Hindu Art of Reckoning*. Today, it has been adopted almost everywhere.

Father of algebra
Drawing on the work of ancient mathematicians, al-Khwarizmi wrote the foundational textbook on algebra. His book gave the first systematic explanation of the symbols and methods used to solve equations. For example, he demonstrated how to work out numerical problems by transposing elements from one side of an equation to the other and canceling out items that appear on both sides. Al-Khwarizmi's work on arithmetic and algebra was built on by later mathematicians and went on to transform the way people did business by simplifying financial transactions and enabling accurate accounting. It also helped scientists hone their theories about the natural world.

One of al-Khwarizmi's achievements was to produce a map of the river Nile. It shows the river flowing from a place called Moon Mountain (right) to the Mediterranean Sea (left).

AL-**KHWARIZMI**

"With my **two algorithms,**
one can solve all problems"

Al-Khwarizmi, c.830 CE

ALHAZEN

An innovator in the field of optics, Alhazen investigated how light is reflected and refracted, and was one of the first experimental scientists.

Born in the Persian city of Basra (in modern-day Iraq) in c.965, Alhazen was educated in Baghdad during the Golden Age of Islamic civilization. He worked as a civil servant and pioneered the now-standard scientific method of proposing hypotheses and then testing them through experiments.

Under house arrest in Egypt, where he had promised to dam the River Nile for the Caliph but failed to do so, Alhazen began his seven-volume *Book of Optics*. He disputed the theories of Euclid and Ptolemy that "rays" are emitted from the eye and bounce back from an object. Instead, he experimented with bulls' eyes and found that light enters a hole (the pupil) and is focused by a lens onto a sensitive surface (the retina) inside the eye. Alhazen also studied the properties of glass, mirrors, and lenses and showed that light travels in straight lines.

MILESTONES

EARLY TEXTS
Writes his first papers in c.1000 on subjects including astronomy, optics, and mathematics.

DOOMED PROJECT
Tasked with regulating the flow of the River Nile in c.1011, but it proves impossible.

FOCUSED STUDY
Resigns from his civil-service job in 1027 to devote himself full-time to scientific study.

"The **seeker after truth** ... submits to **argument** and **demonstration**."

Alhazen, c.1025

HILDEGARD OF BINGEN

A leading authority on medieval pharmacology, Hildegard was a naturalist and healer who explored the medicinal use of plants and animals.

Hildegard was born in Bermersheim, Germany, the youngest child in a family of minor nobility, in 1098. After suffering ill health and seeing visions as a child, she entered a Benedictine abbey at the age of 8. She was made abbess in 1136 and began to study medical treatises from antiquity. She wrote on geology, science, botany, and healthcare, and her work on blood circulation and mental health was advanced for the time. Hildegard prescribed herbs and botanical tonics as both prevention and cure for specific diseases. She believed that ill health was caused by an imbalance in the four humors—blood, yellow bile, black bile, and phlegm.

MILESTONES

NATURAL REMEDIES
In c.1150, writes *Physica*, a nine-volume text listing remedies made from plant and animal extracts.

CURING SICKNESS
Produces major work, *Causae et Curae*, on diseases and their treatments, in c.1150.

MADE SAINT
Beatified by Pope John XXII in 1326; becomes St. Hildegard of Bingen in 2012.

"Gaze at the **beauty** of the **green Earth**. Now, **think.**"

Hildegard of Bingen

Hildegard composed her Book of Divine Works *between 1163 and 1174. It was seen as her most complex work and revealed her visionary approach.*

DIRECTORY

The earliest known scientists appeared in the ancient world, their discoveries immortalized by the advent of writing. Pioneering individuals from flourishing civilizations started devising methods to observe and control nature, cultivating disciplines such as medicine and astronomy in the process.

DEMOCRITUS
c.460–370 BCE

Greek philosopher Democritus, along with his teacher Leucippus, proposed the idea that everything is made up of tiny, unchanging particles called "atoms," which move around freely and combine with other atoms to form different objects. This theory, known as atomism, was the first view of the Universe that did not rely on the notion of a god or gods. Democritus was a modest man who worked hard and acquired fame not for his philosophy, but for his ability to predict changes in the weather. He was known as the "laughing philosopher" for his tendency to laugh at the foolishness of others.

EUCLID
325–c.265 BCE

Often called the "father of geometry," Greek mathematician Euclid is best known for his *Elements of Geometry*, a compilation of the advances made in Greek mathematics over the previous 300 years. *Elements* was translated into Arabic in about 800 CE, and in 1482, it became the first mathematics textbook to be printed. Euclid's principles formed the basis of mathematical teaching in the West and Arab world for more than 2,000 years. Although little is known about his life, Euclid is also said to have written about optics and astronomy.

HIPPARCHUS
190–c.120 BCE

When he was compiling his catalog of stars in 129 BCE, Greek astronomer Hipparchus noticed that the positions of the stars he saw were different from the earlier records he was consulting. He concluded that it was not the stars that had moved, but Earth itself, and so proposed axial precession—a gradual but continuous change in the orientation of Earth's rotational axis. Hipparchus is also said to have invented trigonometry—the branch of mathematics that deals with calculating the angles of triangles and the lengths of their sides.

PLINY THE ELDER
23–79 CE

Pliny the Elder was born in Como, in what is now Italy, and served as a military commander for the Roman Empire. He spent much of his spare time writing and studying, but his only surviving work is *Natural History*, which he completed in 77 CE. Made up of 37 books, *Natural History* covered an unprecedented breadth of ancient knowledge, including botany, astronomy, geography, zoology, and medicine, and became a model for later encyclopedias. Based in the Bay of Naples on his last military assignment, Pliny died after inhaling fumes from the eruption of Mount Vesuvius.

LIU HUI
c.225–298 CE

Little is known about the Chinese mathematician Liu Hui except that he lived in the northern Wei kingdom and wrote two works. The most famous of these was his commentary on *The Nine Chapters on the Mathematical Art*, an early Chinese mathematics handbook that offered mathematical solutions to everyday problems. Liu showed that the methods suggested in the book were correct, and in doing so, he created proofs that validated mathematical methods for establishing the area of circles and the volume of solids such as pyramids, as well as for adding fractions.

BRAHMAGUPTA
597–668 CE

Indian astronomer and mathematician, Brahmagupta grew up in Bhillamala, in northwest India, and went on to become head of the astronomical observatory at Ujjain, in central India. His two major works on astronomy and mathematics were written entirely in verse. Brahmagupta established rules for dealing with positive numbers, which he called "fortunes," and negative numbers, which he called "debts," but his main achievement was his set of rules for the number zero. He showed that when zero is subtracted from or added to a number, that number does not change.

JABIR IBN HAYYAN
721–815 CE

Known in the West as Geber, Abu Musa Jabir ibn Hayyan al-Azdi was born in Tus (now Iran), studied in what is now Yemen, and worked as a physician and an alchemist in what is now Iraq. He is credited with a vast body of work on subjects that ranged from alchemy and astronomy to medicine and magic. Many of his alchemical writings were later translated into Latin and became influential in the evolution of chemistry in Europe. In these writings, ibn Hayyan emphasized the importance of practical experimentation and described many processes still used in modern chemistry, including distillation and crystallization. He died at the age of 94.

ABD AL-RAHMAN AL-SUFI
903–986 CE

Persian astronomer Abd al-Rahman al-Sufi made the first record of what are now known to be galaxies. In *The Book of Fixed Stars*, he described a "little cloud," now known as the Andromeda Galaxy, and also identified the Large Magellanic Cloud, which he called the White Ox. He could not have seen this from his home in Isfahan, in what is now central Iran, so he must have heard about it from astronomers in Yemen and sailors crossing the Arabian Sea. Al-Sufi's main work was translating Ptolemy's *Almagest* into Arabic.

AVICENNA
980–1037

Born near Bukhara, in present-day Uzbekistan, Avicenna (also known as ibn Sina) was a remarkable scholar who had memorized the Qur'an by the age of 10. At 16, he gained fame as a doctor when he cured the Samanid ruler Nuh ibn Mansur of a mystery illness. He was rewarded with the use of the ruler's extensive library, and produced the first of many works soon after. Avicenna traveled widely, working in royal courts across Asia. In philosophy, he is famous for his thought experiment the "Flying Man," with which he argued that the mind is distinct from the body. In 1025, he completed his major work, *The Canon of Medicine*, which explains how to diagnose and treat illnesses.

TROTA OF SALERNO
c.11th century

Trota of Salerno (a city in what is now southern Italy) was a female physician who lived and worked at a time when medical treatments offered to women were often barbaric. Trota approached this problem in a practical way, writing not only about what ought to work in the treatment of women's diseases, but also what did work in practice. Her most influential work was her contribution to the *Trotula*, a popular medical text in medieval Europe.

ROGER BACON
1214–1292

English friar Roger Bacon was a pioneer of experimental science who outlined his ideas about mathematics, physics, optics, and alchemy in his *Opus Majus* ("Great Work"), which he presented to Pope Clement IV as a plea for church reform. Unfortunately, the pope died before he could read the work, and Bacon was imprisoned by his fellow Franciscans for his radical views. His studies of light and how it interacts with objects heralded a new approach in optics. Bacon was also the first European to describe how to make gunpowder. His thirst for knowledge earned him the nickname "Doctor Mirabilis" (Wonderful Teacher).

CHOE MU-SEON
1325–1395

Born into a wealthy family, Choe Mu-Seon's ambition to bring the recipe for gunpowder to Korea was inspired by the fireworks he saw as a boy. Choe eventually achieved his aim by bribing a Chinese merchant to give him the secret recipe. Korea began to produce its own gunpowder soon after in c.1375. Choe went on to invent various firearms, which he put to the test against the pirates who threatened Korea's coastal regions, as well as the Japanese at the Battle of Jinpo in 1380.

ALI QUSHJI
1403–1474

Born in Samarkand, in what is now Uzbekistan, Ali Qushji studied astronomy and mathematics under the ruler and astronomer Ulugh Beg, who founded the Samarkand Observatory (of which Qushji later became director). There, Qushji wrote about the phases of the Moon and contributed to Beg's star catalog. After Beg was killed, Qushji moved to Constantinople (present-day Istanbul), where he founded a school for teaching traditional Islamic sciences. He is best known for his efforts to separate astronomy from philosophy and for providing empirical evidence that the Earth moves.

2

SCIENTIFIC REVOLUTION

1450–1650

LEONARDO
DA VINCI

Beyond being one of the greatest artists ever to have lived, Leonardo da Vinci was a brilliant Renaissance thinker whose work spanned engineering, science, architecture, and mathematics. He believed that observation was a vital tool for inquiry and that art and science were equally important fields of creative knowledge.

MILESTONES

EARLIEST WORK
Paints *The Annunciation* in 1473 while apprenticed to Verrocchio. It is his earliest surviving work.

MOVES TO MILAN
Designs military and engineering projects for the Duke of Milan from 1483.

MILITARY ENGINEER
Works for Cesare Borgia in Rome from 1502, surveying and drawing highly accurate maps.

PERSONAL COLLECTION
Devotes more time to sketching and writing in notebooks following the death of his father in 1504.

ROYAL INVITATION
Invited to the royal court by King François I of France in c.1516; dies there in 1519.

Leonardo da Vinci was born near the town of Vinci in Tuscany, Italy, in 1452. He was educated as the son of a gentleman despite being the illegitimate son of a Florentine legal notary, Piero, and Caterina, a peasant woman. As a teenager, he became the apprentice of Verrocchio, one of the foremost artists of Florence, and learned painting, sculpting, and metalworking, along with mathematics and the sciences.

Possessing a remarkably artistic eye, da Vinci also had an insatiable curiosity for science and mechanics. At the age of 17, he assisted Verrocchio in the construction of a 2-ton (1.8-tonne) gilded copper sphere, which was installed on the summit of Florence cathedral's 376-ft- (114-m-) high dome in 1469. He was fascinated by the mechanical working of the hoists used to lift the sphere to such a height.

Human mechanics
In keeping with the artistic education of the Renaissance period, da Vinci received a grounding in anatomy during his apprenticeship and studied the skeletal and muscular structure of the human figure. As revealed by the contents of his notebooks—many of which were

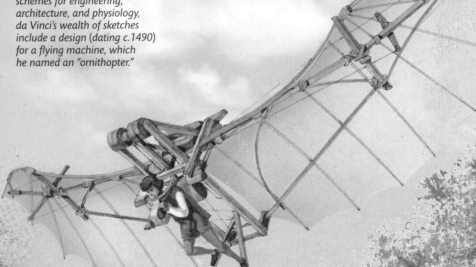

As well as practical, applied schemes for engineering, architecture, and physiology, da Vinci's wealth of sketches include a design (dating c.1490) for a flying machine, which he named an "ornithopter."

compiled later in his career but show subjects of earlier interest—anatomy became one of his primary concerns. It is likely that da Vinci began conducting his own dissections after 1483, when he entered the employ of Duke Ludovico Sforza in Milan, which was a leading center of medical investigation. He later attested to having performed some 30 dissections during his career.

Knowing how to see

Da Vinci believed that painting itself was a science, primarily because of art's capability of being, in the right hands, "the sole imitator of all the manifest works of nature." Observation was the essential tool of this discipline, which da Vinci summed up in his maxim *saper vedere*—which means "knowing how to see" one's subject. Unlike sculpting, drawing and painting require the artist to render a three-dimensional scene or object in two dimensions,

> "He who **loves practice without theory** is like a sailor **with no rudder.**"

Leonardo da Vinci

meaning that artists of these media must achieve mastery of mathematical and optical principles.

A devotee of the mathematical basis of his art, da Vinci believed that geometry and proportion were essential to achieve perspective. He was particularly interested in

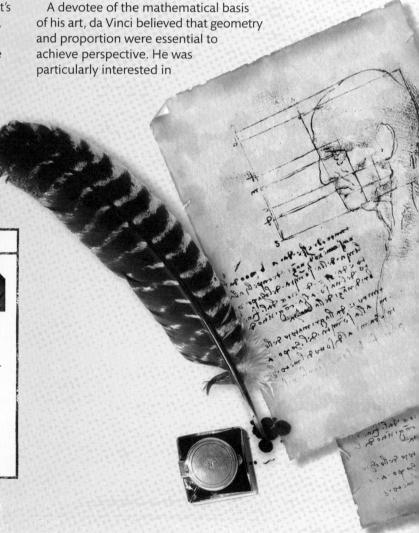

LEONARDO **FIBONACCI**

Italian mathematician Leonardo Fibonacci popularized Hindu–Arabic numbers.

Born in Pisa, Fibonacci (c.1170–1250) learned the Hindu–Arabic numeral system while helping his merchant father in Algeria. He then studied with Arab mathematicians, writing up his findings in *Liber Abaci* and *Practica Geometriae*. He also devised the Fibonacci sequence, in which each number is the sum of the two preceding it: 0, 1, 1, 2, 3, 5, 8, and so on. Squares of each value can be drawn and aligned to create a spiral.

the work of the mathematician Leonardo Fibonacci, who drew links between numerical sequences and geometrical shapes (see box).

Enhanced accuracy

From 1502, da Vinci worked for the statesman Cesare Borgia in Rome. He completed a series of cartographic surveys, in which he used an aerial perspective rather than the traditional oblique view. He paced out the roads and squares to measure distances and used the newly invented theolodite—an optical device used by surveyors—to measure angles, achieving an accuracy that had never been seen before.

Renowned as a brilliant thinker and artist beyond comparison, da Vinci was invited to France by King François I in 1516. He lived out his final years free from any obligation to paint, and instead prepared his many papers—on everything from anatomy to hydraulic engineering—for publication. However, due to the vast scale of his creativity, it proved too immense a task for da Vinci himself to complete.

Da Vinci's notebooks are full of sketches of the human body. He had studied anatomy as part of his training as an artist, but by 1490, it had become a distinct research area for him.

DISSECTED **30** CORPSES *TO STUDY HUMAN ANATOMY*

PERFECTED HIS OWN **MIRROR SCRIPT** FOR SECRECY

13,000 PAGES OF **PEN** AND **INK** NOTES AND SKETCHES SURVIVED

In the 16th century, Polish astronomer and mathematician Nicolaus Copernicus challenged the prevailing view of the world by suggesting that the Sun, rather than the Earth, is at the center of the Universe.

Born in Poland in 1473, Nicolaus Copernicus studied astronomy and astrology in Kraków before traveling to Italy to study canon law (laws relating to the Church) at the University of Bologna. He later studied medicine at the University of Padua. After gaining a doctorate in canon law from the University of Ferrara in 1503, he moved to Wrocław, in western Poland, to devote his time to his greatest passion: astronomy. During these years, Copernicus formulated a cosmological theory that broke with centuries of tradition and eventually redefined the science of astronomy.

The established Universe
At the time Copernicus was working, there was a single accepted view of how the Universe was organized—the geocentric model. This model placed the Earth statically at the center of the Universe, with the planets, the Sun, the Moon, and the stars revolving around it. This view, proposed by Aristotle (see pp.14–17) and refined by Ptolemy (see p.33), had dominated cosmology for centuries. It was also entrenched within the doctrine of the Catholic Church, which considered humans to be the pinnacle of God's creation and the Earth to be at the center of everything. However, there were serious problems with the geocentric model. Chief among these was its failure to

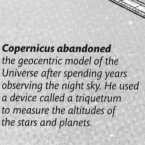

Copernicus abandoned the geocentric model of the Universe after spending years observing the night sky. He used a device called a triquetrum to measure the altitudes of the stars and planets.

NICOLAUS

COPERNICUS

account for the fact that many planets appear to turn back in their orbits, performing loops in the sky that Ptolemy called "epicycles," or orbits within orbits. Copernicus set out to explain this, and his answer showed that the planets do not turn back in their orbits, but only appear to do so as the Earth overtakes them in their journey around the Sun.

Breaking with convention

Copernicus had gained a reputation as a highly adept astronomer, and in 1514, he privately circulated copies of a handwritten pamphlet that proposed an alternative theory of planetary motion. He placed the Sun at the center of the Universe, with the Earth and all the planets orbiting it in a fixed order. Further, he argued that the Earth spins on its axis, accounting for the observed daily movements of the planets, the stars, the Moon, and the Sun. Finally, he proposed that the stars lie far beyond the farthest planet, suggesting that the Universe must be much larger than had previously been believed.

This revelatory view of the solar system solved many—but not all—of the problems with the geocentric model. Copernicus himself had trouble with it and was forced to follow Ptolemy's example and include epicycles in order to account for certain phenomena. (He wrongly believed that the planets move in circular rather than elliptical orbits.)

However, despite its imperfections, Copernicus's Sun-centered, or heliocentric, model of the Universe (*helios* meaning "Sun" in Greek) greatly outstripped the Ptolemaic model in terms of accuracy and gradually gained credence in intellectual circles. Georg Joachim Rheticus, a German

mathematician and pupil of Copernicus, strongly encouraged him to publish his theory, but he only did so in 1543, as he was dying. His seminal work was published as *De Revolutionibus Orbium Coelestium* (*On the Revolutions of the Celestial Spheres*). It contradicted the Church's teachings but was only branded heretical in 1616, when it was endorsed by Italian astronomer Galileo Galilei (see pp.50–55).

Posthumous influence

Providing a radical new view of the Universe and challenging centuries of received wisdom, Copernicus's controversial work paved the way for the Scientific Revolution and the rebirth of philosophy in Europe. It marked the end of an old idea— that humans stand at the center of things—and set up a dispute between science and religion that continues to this day.

GIORDANO **BRUNO**

An Italian friar and astronomer, Bruno was executed for supporting the heliocentric model.

Part of the Dominican order, Bruno (1548–1600) was a radical thinker who rejected the Church-endorsed geocentric view of the Universe. Instead, he supported Copernicus's heliocentric model and took the theory further by suggesting the Universe was infinite and might have numerous worlds within it. His unorthodox views angered Church authorities, who tried him for heresy. He was burned at the stake in 1593.

> **"**In the **middle** of everything is the **Sun."**

Nicolaus Copernicus, 1543

WAITED
29
YEARS
BEFORE
PUBLISHING
HIS THEORY

ONLY **27**
OF HIS
RECORDED
OBSERVATIONS
HAVE **SURVIVED**

Like Ptolemy, Copernicus believed the planets were embedded in crystalline spheres that orbited in a fixed circular motion.

GALILEO
GALILEI

A scientist of many talents, Galileo Galilei made key discoveries that transformed our view of our place in the Universe. He improved the telescope, showed that the Sun was the center of the solar system, and revolutionized theories about the forces of motion, which had been unchanged since Aristotle's time.

Born in Pisa, Italy, Galileo Galilei was the son of a musician and the eldest of six children. He initially wanted to become a monk, but his father wanted him to be a doctor, so he went to study medicine at the University of Pisa. However, while there, he spent most of his time studying mathematics, which he found much more interesting. He went on to become professor of mathematics at Pisa at the age of 25, then later at the University of Padua.

During his time as a student, he watched the big bronze lamp in Pisa Cathedral swinging back and forth. Galileo timed the swings using his pulse and noticed that each swing, however long or short, took the same time. Realizing the pendulum is a regular timekeeper, he invented a simple device for measuring pulses: a small pendulum with a length of string that could be adjusted so that the swing matched the speed of the person's pulse.

Galileo began to conduct experiments to find out why objects move or stop moving. At the time, scientists still believed what Aristotle had said 2,000

MILESTONES

CREATES TELESCOPE
Designs and builds his own telescope in 1609, which reveals the Moon and planets in unprecedented detail.

ASTRONOMICAL FINDS
In 1610, discovers the moons of Jupiter and writes about his observations in *The Starry Messenger*.

TRIED FOR HERESY
Ordered by the Church in 1616 to retract his claim that the Earth is not the center of the Universe.

HOUSE ARREST
Tried for heresy again by the Inquisition in 1633. Spends rest of his life under house arrest.

Galileo was accused of heresy for his belief that the planets orbit the Sun. But in 1992, three and a half centuries later, the Pope officially admitted that Galileo had been right.

years earlier: that objects move because they are propelled by an external force. No one had challenged this because it seemed common sense. But "common sense" was not enough for Galileo, who believed that theories had to be tested by observation and precise measurement.

Gazing at the stars

In 1609, Galileo became interested in a new invention: the telescope (see box). This changed the course of his research dramatically. He set about building his own improved version, with stronger magnification. With it, he made some surprising discoveries. He saw that the Moon had mountains and craters and that, like the Moon, Venus changed shape in crescent phases. Most startling was his discovery that

" The **passage of time** has **revealed to everyone** the **truths** that I previously set forth. **"**

Galileo Galilei, 1615

Jupiter had its own moons. At the time, it was accepted that the Earth was the center of the Universe, with the Sun, Moon, stars, and planets orbiting around it; that Jupiter's moons were clearly orbiting Jupiter shook this concept to its core. Galileo became convinced that the planets orbit the Sun—a theory that was first proposed by Nicolaus Copernicus in 1543.

Galileo was the first person to use lenses to study the solar system. Jupiter's four largest moons—Io, Europa, Callisto, and Ganymede—are named the "Galilean moons" in honor of his discovery in 1610.

However, this claim was not popular with the Catholic Church. In 1616, Galileo was brought before the Inquisition (the Church court) and told to take back what he had said. Grudgingly, he did so. However, he went on to publish his ideas, and in 1633, he was tried again. This time, he was placed under house arrest, where he remained until he died. During that time, he returned to his work on motion. He realized that a moving object will keep moving at the same speed unless a force (such as friction) acts on it to slow it down or speed it up—this is the principle of inertia. He concluded that in a vacuum, all objects will fall at the same speed, no matter their size and weight. This revolutionized the basic laws of physics and provided the foundation for Isaac Newton's three laws of motion, which are fundamental to our understanding of the Universe.

HANS **LIPPERSHEY**

A Dutchman who made spectacles, Hans Lippershey was one of the first people to build an instrument for viewing distant objects—a telescope—but he failed to secure a patent for it.

In 1608, Lippershey (1570–1619) applied for a patent for his instrument that could magnify objects to three times the size. According to one story, he got the idea for his "looker" from two children who were playing in his shop, putting one lens in front of another. However, some said that he stole the concept from another spectacle-maker, Zacharias Janssen. Furthermore, a similar patent was filed just a few weeks after Lippershey's by Jacob Metius. Due to the dispute, the Dutch government refused both patent applications but paid Lippershey well for his design. When Galileo heard about the telescope, he made his own.

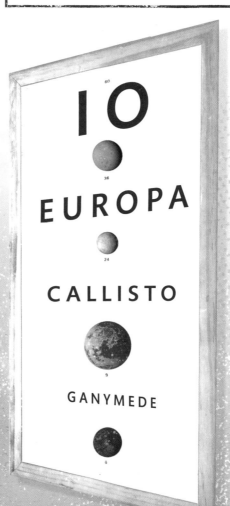

IO

EUROPA

CALLISTO

GANYMEDE

HIS **TELESCOPE MAGNIFIED** NORMAL VISION BY A FACTOR **OF 20**

CONCEIVED THE FIRST PENDULUM CLOCK

"PHILOSOPHY IS WRITTEN IN THAT GREAT BOOK WHICH EVER LIES BEFORE OUR EYES—I MEAN THE UNIVERSE—BUT WE CANNOT UNDERSTAND IT IF WE DO NOT FIRST LEARN THE LANGUAGE AND GRASP THE SYMBOLS, IN WHICH IT IS WRITTEN."

Galileo Galilei
The Assayer, 1623

*This artwork from a 1974 issue of **National Geographic** magazine*
illustrates the Church's initial acceptance of Galileo's observations. ▶

JOHANNES KEPLER

Johannes Kepler was a German astronomer and mathematician who formulated three laws of planetary motion from observed astronomical data. In providing evidence to support a heliocentric (Sun-centered) Universe, he laid the foundations of modern astronomy and also advanced the study of optics.

A gifted mathematician, in 1589, Johannes Kepler earned a scholarship to the University of Tübingen, where he was tutored by the leading German astronomer Michael Mästlin. Although the university taught geocentric astronomy (with planets revolving around the Earth), Mästlin introduced Kepler to the controversial idea of a heliocentric Universe, first proposed in 1543 by Nicolaus Copernicus (see pp.46–49).

Astronomy and astrology

In 1594, Kepler began teaching mathematics in Graz, Austria, but his spare time was devoted to studying the planets. During the late 1500s, astronomy was of less interest to the wider public than the astrological art of casting horoscopes based on predicted planetary movements. Astronomers such as Kepler used astrology as their chief source of income and influence, and therefore the accuracy of their predictions was of great significance to their livelihoods.

In 1600, Kepler was invited by Tycho Brahe, a Danish astronomer (see p.58), to be his assistant. Brahe made the most accurate observations of

Kepler made a model of the Universe using five polyhedra (three-dimensional objects with polygon faces). Each was set within a sphere and arranged in concentric sizes.

MILESTONES

HELIOCENTRIC SUPPORT
Produces a written defense of the Copernican model of the Universe in 1596, risking displeasure of the Church.

INFLUENTIAL POSITION
Becomes assistant to Tycho Brahe in 1600; appointed Imperial Mathematician to Rudolph II a year later.

OPTICAL WORK
Begins study of optical theory in 1603; publishes *Astronomiae Pars Optica*, his landmark text, in 1604.

PLANETARY LAWS
Proposes his first two laws of planetary motion in 1609; works for 10 years before publishing his third law.

> **"**Geometry is **one and eternal**, a reflection of the **mind** of God.**"**

Johannes Kepler, 1610

57

CALCULATED THAT MARS HAD AN ELLIPTICAL ORBIT AFTER

40

FAILED ATTEMPTS

HIS **3RD LAW** TOOK HIM

10

YEARS TO FORMULATE

DISCOVERED SATURN TAKES ROUGHLY

29 EARTH YEARS

TO ORBIT THE SUN

TYCHO **BRAHE**

Danish astronomer Tycho Brahe accurately cataloged the stars before the telescope.

In 1577, Brahe (1546–1601) observed a new star in the constellation of Cassiopeia and determined to accurately record celestial bodies. He built a set of instruments with which to measure the positions of the stars and spent 20 years compiling data. The accuracy of his observations surpassed those of any other practicing astronomer and enabled his assistant, Johannes Kepler, to calculate his laws of planetary motion.

the planets and asked Kepler, who had the better mathematical skills, to investigate the complicated orbit of Mars.

Laws of planetary motion

When Brahe died in 1601, Kepler took over his role as Imperial Mathematician to Rudolph II, Holy Roman Emperor. Using Brahe's extensive observations, compiled over 20 years, Kepler resumed his study of Mars, which resulted in his three laws of planetary motion.

Unable to match Brahe's observational data to Copernicus's criteria—that each planet has a circular orbit and moves at a constant velocity—Kepler adjusted his calculations. He began experimenting with different orbital shapes, eventually concluding that a planet's orbit must be elliptical (like a stretched circle) and that the extent (or eccentricity) of that ellipse would vary from planet to planet. This was Kepler's first law of planetary motion and successfully explained the apparent retrograde (backward) motion of Mars as viewed from Earth.

His second law stated that a planet's velocity was not constant and that the closer it came to the Sun, the faster it would travel. These laws, published in his 1609 text *Astronomia Nova*, represented a landmark in astronomical history.

Published a decade later, Kepler's third law of motion mathematically described the relationship between a planet's year (orbital period) and its average distance from the Sun. He calculated the orbital period for each planet and in comparing them realized that if a planet is twice as far from the Sun as another planet, then its orbital period will be three times longer than that of the nearer planet.

Influences and achievements

The accuracy of Kepler's laws gave credence to the heliocentric view of the Universe and ensured its wide acceptance. His work in astronomy gave Newton the basis on which he would later devise his universal laws of gravitation and marked a profound shift in scientific thought.

While he is best known for his astronomical work, Kepler also made great advances in the study of optics and produced one of the first key texts in this field. With achievements that included explaining the visual mechanics of depth perception and proposing designs for eyewear to combat near- and farsightedness, he is also considered a pioneer in the field of modern optics.

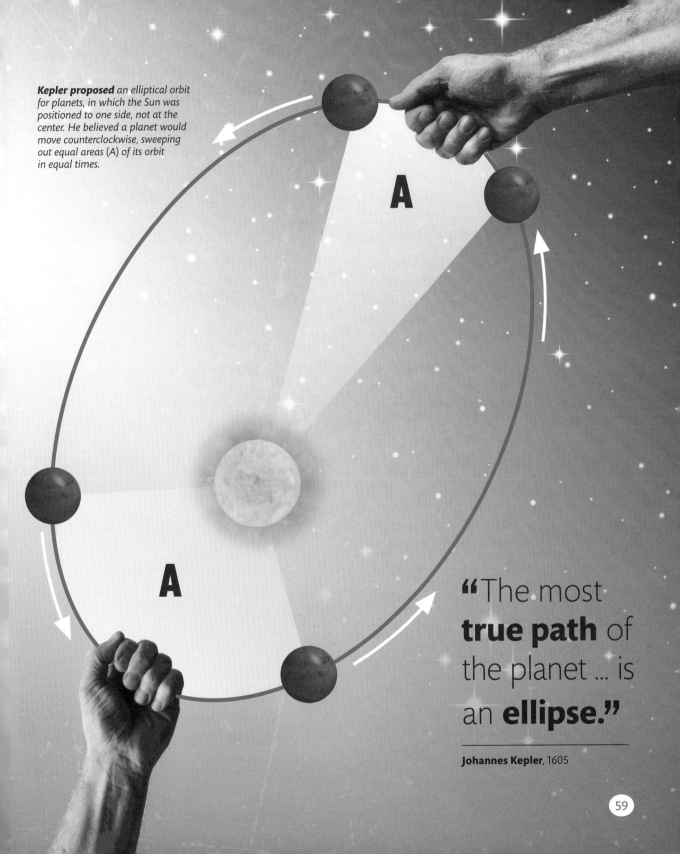

Kepler proposed an elliptical orbit for planets, in which the Sun was positioned to one side, not at the center. He believed a planet would move counterclockwise, sweeping out equal areas (A) of its orbit in equal times.

A

A

"The most **true path** of the planet ... is an **ellipse.**"

Johannes Kepler, 1605

British physician and natural historian William Harvey was the first person to understand how blood circulates around the body and to correctly explain the role the heart plays in this process. His work replaced centuries-old speculation with a theory based on his research into the circulatory systems of animals.

William Harvey was the son of a wealthy merchant. He studied medicine at Cambridge University and then at the University of Padua, Italy. His teacher at Padua, the esteemed anatomist Hieronymus Fabricius, was a great influence on him. Fabricius had discovered that human veins had one-way valves, although he did not understand why.

On his return to London, Harvey became a lecturer in surgery. His reputation grew, and he was appointed Royal Physician. But his main interest was in research. He was dissatisfied with the prevailing view of circulation, which dated back to the ancient Greek medical practitioner Galen (see pp.24–29), who believed that blood was made in the liver and was sucked from the veins by the heart.

Harvey believed that experimentation was crucial to medicine and carried out many dissections—not only of dead bodies but also of living animals. Although the latter was brutal by modern standards, it led to his discovery that the heart is a pump and that it drives the circulation of the blood. He described this system in detail and accurately predicted the existence of capillaries, which transfer blood from the arteries to the veins, thereby completing its circulation around the body.

MILESTONES

RADICAL THEORY
Tells the Royal College of Physicians, at the 1616 Lumleian lecture, that the heart drives circulation.

ROYAL DOCTOR
Becomes court physician to King James I in 1618, and then to King Charles I in 1632.

SEMINAL TEXT
Publishes *De Motu Cordis*, his text on blood and the heart, in 1628, dedicating it to King Charles I.

"The blood is driven into a round ... **and it moves perpetually."**

William Harvey, 1628

Harvey explained his theory of circulation in his key work De Motu Cordis (On the Motion of the Heart). *In it, he illustrated how one-way valves in the veins ensure that blood flows in only one direction.*

DIRECTORY

The spread of rediscovered classical and Islamic texts throughout the 1500s incited an intellectual revolution across Europe. Defiant thinkers, championing observation and testable theories, modernized the scientific mindset with their new understanding of knowledge based on empiricism.

JACOPO BERENGARIO DA CARPI
1460–1530

Italian physician Jacopo Berengario da Carpi was one of the first to use drawings from nature to illustrate an anatomy book and the first to describe heart valves. The son of a surgeon, he was fascinated by human anatomy and claimed to have dissected several hundred bodies. Berengario studied medicine in Bologna, and went on to teach at the university there. He gained a reputation across Italy for his skill and declined a request to become the pope's doctor. He died a wealthy man.

PARACELSUS
1493–1541

Philippus Aureolus Theophrastus Bombastus von Hohenheim, who called himself Paracelsus (meaning "above or beyond Celsus," a Roman scholar), was a Swiss physician and alchemist who established the role of chemistry in medical practice. Inspired by the folk medicine he learned about on his travels, Paracelsus advocated studying nature and using experiments to learn about the human body. He believed that disease was caused by external influences and that it could be treated with external substances, given in the correct dose. He thought that a goiter was caused by minerals that are found in drinking water and introduced the use of mercury as a cure for syphilis (now obsolete). Most modern vaccines stem from Paracelsus's belief that that which makes a man ill can also cure him.

MICHAEL SERVETUS
1509–1553

Spanish theologian and physician Michael Servetus discovered pulmonary circulation (that aeration of the blood occurs in the lungs and not, as Claudius Galen thought, in the heart) almost incidentally as part of his theological quest to understand the soul, which he claimed could be found in the blood. Both Catholics and Protestants deemed his religious views heretical, and he was burned at the stake in Geneva, Switzerland, in 1553, along with almost every copy of his book.

ANDREAS VESALIUS
1514–1564

Flemish-born doctor Andreas Vesalius studied medicine at the University of Padua, Italy, and became professor of surgery there aged 23. Anatomists of the time relied not on dissection, but on the work of Greek and Roman scholars such as Claudius Galen, whose knowledge of anatomy rested on examining animals. Vesalius challenged this tradition, dissecting human bodies in front of his students and creating an accurate record of what he saw in his 1543 masterpiece *De Humani Corporis Fabrica* (*On the Fabric of the Human Body*). He was physician to the Holy Roman Emperor Charles V, then to Philip II. He died in obscurity on a Greek island following a shipwreck.

GABRIELE FALLOPPIO
1523–1562

Due to lack of funds, Gabriele Falloppio initially joined the clergy, but when his situation improved, he went to study medicine instead. The Italian's skill at dissection earned him the position of professor of anatomy at Pisa in 1548, and he took over from Andreas Vesalius at the University of Padua 3 years later. Falloppio is best known for his study of human reproductive organs and for describing the narrow tubes that connect the ovaries to the uterus (fallopian tubes). He also studied syphilis and designed the first condom as a means of preventing spread of the disease.

WILLIAM GILBERT
1544–1603

After years of painstaking experiments, William Gilbert concluded that the Earth was magnetic. The English physician studied ships' compasses, then he made a model globe out of lodestone (a magnetic rock) and used it to test compass needles. He concluded that Earth is a giant magnet and that its north and south magnetic poles more or less correspond with its geographic ones. Gilbert was Queen Elizabeth I's physician for the last 2 years of her reign. He died soon after the queen, probably from the plague.

GIORDANO BRUNO
1548–1600

Burned alive as a heretic, Giordano Bruno believed that the Universe had no center and that there are many worlds, and many suns, just like ours. The Italian philosopher, mathematician, and one-time monk made the Sun seem insignificant with his discovery that it is just one star surrounded by planets and moons, among an infinite number of others. In doing so, he rejected the traditional Earth-centered cosmology, went further than the Sun-centered one described by Nicolaus Copernicus, and came up with a more accurate picture of an infinite Universe.

JOHN NAPIER
1550–1617

Born near Edinburgh, Scotland, John Napier joined St. Andrew's University in 1563, but left a year later to continue his mathematical studies in Europe. On his return, he wrote an interpretation of the Bible's Book of Revelation and designed various war weapons. But Napier's best-known achievement is his invention of the logarithm, a number that represents another number in order to simplify a difficult division or multiplication sum. He also created "Napier's bones," a set of rods that could be used to multiply or divide numbers.

FRANCIS BACON
1561–1626

After studying law, the Englishman Francis Bacon became a member of parliament in 1581. He was unpopular with Queen Elizabeth I, but found favor with James I and rose to become Lord Chancellor in 1618. Two years later, he was convicted of bribery and banned from public office. Undeterred, he pursued his interest in science and published Novum Organum (New Instrument) in 1620. Denouncing Aristotle's approach as unscientific, Bacon outlined his scientific method: make an observation, develop a theory that might explain that observation, and test it using experiments. It has been the mainstay of science ever since.

WILLEBRORD SNELL
1580–1626

Astronomer and mathematician Willebrord Snell was born in Leiden, the Netherlands, and succeeded his father as professor of mathematics at the university there. Although he wrote on astronomy, navigation, and geodesy (ways of measuring the Earth), he is best known for discovering the law of refraction in 1621. When light passes through air, it travels in a straight line; but when it passes through denser materials, it changes direction—that is, it is refracted. Snell's law relates how much the light bends to the properties of the material it is passing through.

PIERRE GASSENDI
1592–1655

Philosopher, mathematician, and priest Pierre Gassendi criticized his fellow Frenchman René Descartes' description of a mechanical Universe filled with particles of matter and proposed an alternative view that was more compatible with his Christian beliefs. He suggested that atoms share some of the properties of the physical things they make up, such as shape and size, and that atoms may join together to make larger molecules. Unlike Descartes, he believed in the existence of vacuums, going as far as to say that most matter consisted of "void."

EVANGELISTA TORRICELLI
1608–1647

After studying in Rome, Italian physicist and mathematician Evangelista Torricelli worked as an assistant to Galileo Galilei for the last 3 months of the astronomer's life. At Galileo's suggestion, Torricelli filled a tube with mercury and upended it into a bowl of mercury. When some of the mercury stayed in the tube, he realized that the space above it was a vacuum and that the height of the mercury in the tube was determined by atmospheric pressure. He had created the first barometer. A unit of pressure, the torr, is named after him.

MARIA CUNITZ
1610–1664

Growing up in Schweidnitz, Silesia (now Swidnica, Poland), Maria Cunitz had no formal education but was taught by her father, a doctor. Her second husband, Elias von Löwen, also a doctor, taught her mathematics and medicine and encouraged her interest in astronomy. The couple fled Silesia during the Thirty Years' War, and Cunitz spent her time in exile working on a set of astronomical tables that simplified Johannes Kepler's Rudolphine Tables (1627). She published her work in the book Urania Propitia. A fire destroyed Cunitz's home in 1656, and most of her papers were lost.

3

REASON AND ENLIGHTENMENT

1650–1800

Anglo-Irish natural philosopher, theologian, and writer Robert Boyle carried out innovative experiments on the nature of gases. Widely considered the first modern chemist, he redefined chemistry as a discipline and broke away from the Aristotelian methods that had previously governed scientific endeavor.

Born into a wealthy Irish aristocratic family, Robert Boyle was educated at Eton College and continued his studies during an educational tour of Europe. In Italy, he witnessed the reactions to the death of astronomer Galileo Galilei (see pp.50–55) and became greatly inspired by the idea of scientific discovery. In 1644, he moved to England and, although a devoutly religious man, pursued a career in experimental science that epitomized the 17th century's shift towards rationalism.

New era

During the early 1600s, scientific study was typically based on the principles outlined by Aristotle 2,000 years before (see pp.14–17). However, a new era was dawning in which theories were being rigorously tested by experimentation. In London, Boyle met a group of intellectuals who shared his passion for science, which at the time was known as "natural philosophy." This group, called the Experimental Philosophy Club, bought all the equipment they needed to perform experiments and shared their results with

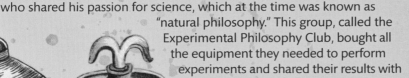

To test whether sound can pass through a vacuum, Boyle placed a bell in a container and drew the air out. As the vacuum increased, the sound of the bell grew fainter.

MILESTONES

NEW PHILOSOPHY
Settles in England in 1644, and turns to experimental science; becomes the leading figure in the field.

AIR PRESSURE
Designs an air pump with Robert Hooke in 1659; pens his first scientific paper on air pressure in 1660.

UNIVERSAL MATTER
Publishes *The Sceptical Chymist* in 1661, which rejects Aristotle's account of the elements.

BOYLE'S LAW
Establishes the universal law of air pressure (the first gas law), publishing his results in 1662.

"Even when we find **not** what **we seek**, we find **something else**."

Robert Boyle, 1661

"There is a **spring**, or **elastical power** in the air we live in."

Robert Boyle, 1660

each other. In 1660, several members, including Boyle, founded the Royal Society, the world's first scientific body.

Experimental science

Using income from his estate to fund his endeavors, Boyle devoted himself to scientific research. With his assistant Robert Hooke (see pp.70–75), he devised an air pump, or vacuum chamber, which enabled him to study the physical properties of air. His experiments demonstrated that sound cannot travel through a vacuum, but that magnetic forces and light can. He also showed that combustion cannot take place in a vacuum.

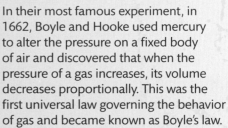

JOSEPH LOUIS **GAY-LUSSAC**

French chemist and physicist Joseph Louis Gay-Lussac was an early investigator into the behavior of gases; he established several laws that describe their behavior.

An accomplished scholar, Gay-Lussac (1778–1850) specialized in the behavior of gases. In 1802, he provided evidence to prove Charles's law, which said that gases expand by the same amount with the same rise in temperature. This was the first gas law since Boyle's law. He also established Gay-Lussac's law, which states that the pressure of a given mass of gas varies directly with the absolute temperature of the gas, when its volume is kept constant. Gay-Lussac's many investigations involved two ascents in a hydrogen balloon, one of which was a solo flight that reached a height of 23,000 ft (7,016 m).

In their most famous experiment, in 1662, Boyle and Hooke used mercury to alter the pressure on a fixed body of air and discovered that when the pressure of a gas increases, its volume decreases proportionally. This was the first universal law governing the behavior of gas and became known as Boyle's law.

Boyle successfully distinguished chemistry from the mystical practice of alchemy and established chemistry as a science in its own right. He rejected the old Aristotelian elements of air, fire, earth, and water and provided the first modern definition of an element as a substance that cannot be broken down into simpler constituents. He also argued that air was composed of tiny, moving "corpuscles" of matter, which paved the way for the development of modern kinetic theory. Like Galileo, Boyle believed that the natural world is governed by mathematics and wanted science to be a process of empirical investigation. He meticulously recorded every stage of his investigations and in doing so became a pioneer in modern experimental methods.

Boyle's work was at the heart of an intellectual movement that saw the development of new theories about the natural world. An innovative thinker who led the charge in experimental science, he achieved both national and international renown during his lifetime.

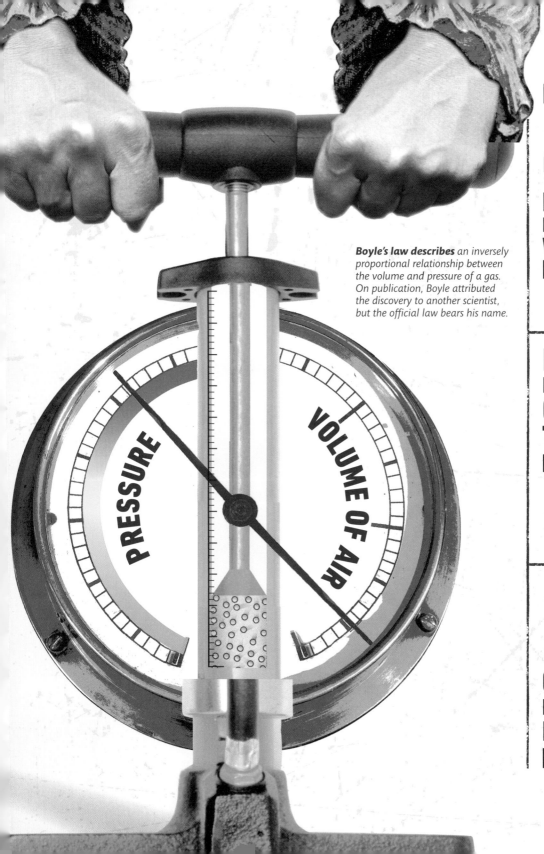

Boyle's law describes an inversely proportional relationship between the volume and pressure of a gas. On publication, Boyle attributed the discovery to another scientist, but the official law bears his name.

PRESSURE

VOLUME OF AIR

RECORDED

43

DIFFERENT EXPERIMENTS WITH A PUMP **IN 1 BOOK**

BELIEVED BASE METALS **COULD BE TURNED INTO GOLD**

21

OF HIS **24** PREDICTIONS HAVE BEEN **REALIZED**

A British 17th-century scientist at the forefront of the Scientific Revolution, Robert Hooke excelled in various roles, including physicist, biologist, inventor, and architect. He discovered the law of elasticity, invented a new microscope, helped to redesign London after the Great Fire, and published the first popular science book.

Robert Hooke was born on the Isle of Wight in 1635. He was often sick as a child, resulting in a sporadic formal education; instead, he engaged in varied pursuits, including painting, model-making, and music. He was highly intelligent and took a great interest in mechanical devices—he famously dismantled a brass clock, then built a working replica out of wood. In 1648, Hooke's father died, and with his inheritance, Hooke funded his own education at Westminster School; aged 18, he gained a place at Oxford University. Hooke's formative education was unorthodox and eclectic, but it formed a springboard that would launch the career of one of the great polymaths of the 17th century. Despite his renown, no contemporary portraits of Hooke survive, but several artists have made posthumous ones.

Wonderful world of science
A lack of funds at Oxford led Hooke to take on part-time work, first as assistant to the physician and chemist Thomas Willis, then later to the natural philosopher Robert Boyle (see pp.66–69). At this time, Hooke's own natural aptitude for science began to flourish and set him on a path toward great future scientific endeavor. Boyle

Hooke discovered the law of elasticity: that the extension and compression of a spring is proportional to the force applied to it. He first published his law as an anagram, which unscrambled read "as the extension, so the force" in Latin.

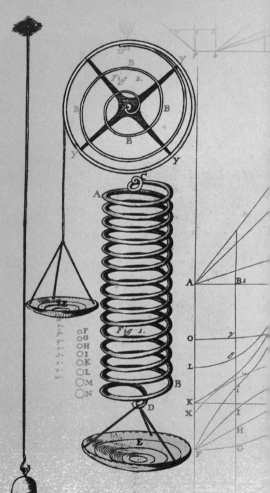

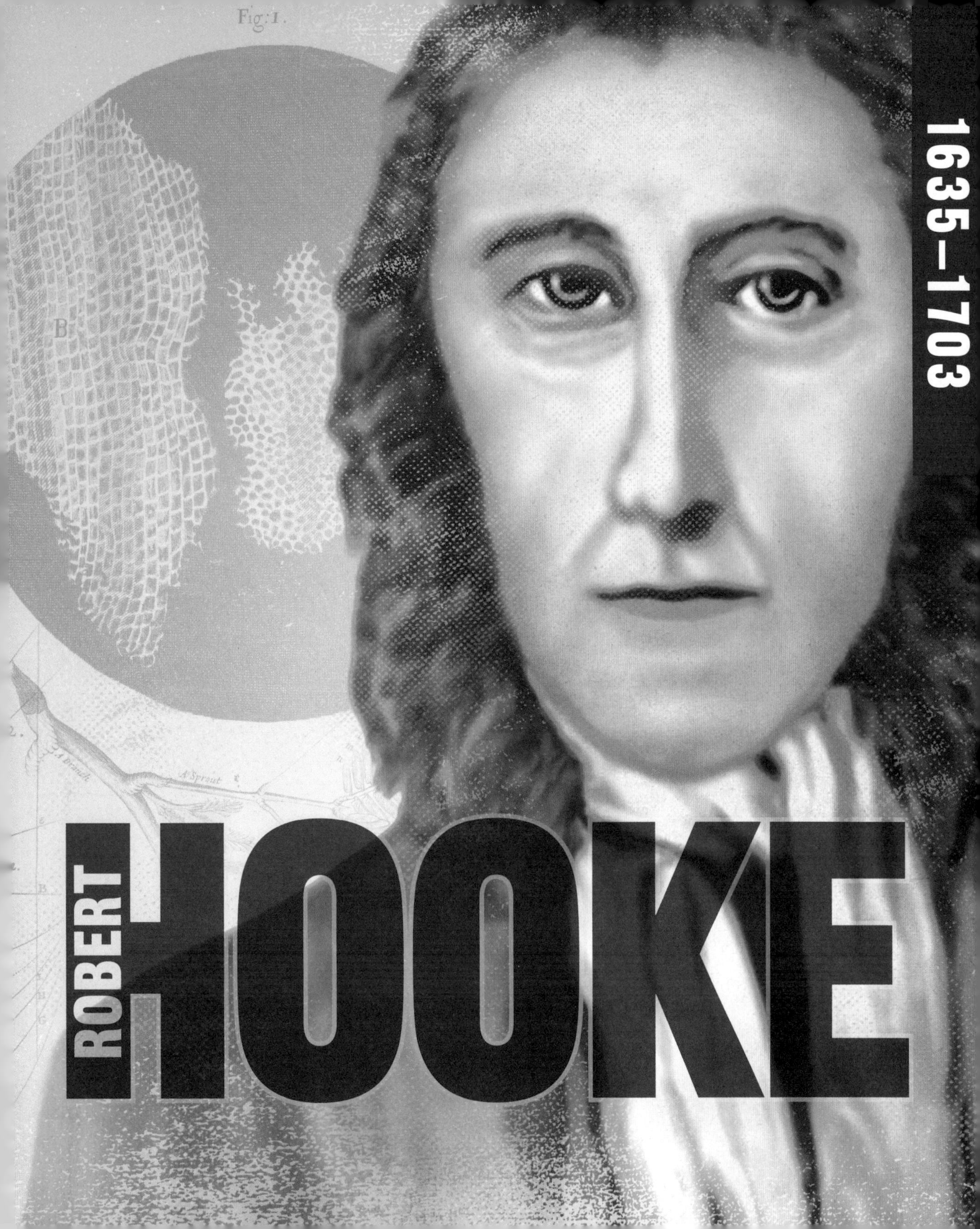

Fig: 1.

B.

A Sprout

ROBERT HOOKE

treated Hooke as an equal, and they worked as partners on the design and build of the first air pump during their investigations into vacuums. These led them to the discovery of Boyle's law on gas compression, yet Boyle himself acknowledged Hooke's contribution.

While working with Boyle, Hooke also carried out many experiments of his own. In 1657, he significantly improved the pendulum clock by inventing the anchor escapement—a vital mechanism that maintains the pendulum's regularity. He also studied the elasticity of springs and applied his findings to the art of clockmaking, inventing the balance-spring for pocket watches, which enabled accurate timekeeping. In 1660, he formulated his law of elasticity, later known as Hooke's law, which states that the tension in a spring increases in direct proportion to the force applied to it.

Curator of experiments

Hooke became part of a circle of like-minded intellectuals who engaged in regular meetings aimed at scientific advancement. In 1660, after Charles II was restored to the throne, science began to thrive again and the group, which included members such as Boyle and Christopher Wren, established themselves as the Royal Society. Hooke was appointed as Curator of Experiments. In this role, he set up and demonstrated experiments either of his own devising or proposed by members, which ranged from investigating the nature of air to gravity and barometric pressure. Hooke's immense skill and aptitude earned him fellowship to the Royal Society in 1663.

In this capacity, Hooke devoted himself to scientific research on a wide, diverse scale, including biology, astronomy, the

"The most ingenious book I ever read in my life."

Samuel Pepys, 1665

nature of gravity, and optics. His abilities as an inventor and mechanic enabled him to personally carry out necessary improvements to the equipment he used, as well as to design new apparatus. Most notably, he built a compound microscope with a new screw-based mechanism to enable the viewer to focus; he also devised a way of illuminating the specimen beneath the lens. Hooke's new microscope enabled him to produce *Micrographia* in 1665—a full study of the microscopic natural world, which he observed, recorded, and illustrated in unprecedented detail. Through intricate, large-scale line drawings, Hooke revealed the structure of minute organisms, such as a flea, with an accuracy never before seen by the human eye, and in doing so produced the world's first best-selling science book. Hooke coined the term "cell," which he used in the book, and laid the foundations for the field of microbiology.

Fire and fortune

After the Great Fire of London in 1666, Hooke assisted Christopher Wren in the architectural redesign of the capital. Although Wren largely took the glory for this work, Hooke designed a significant proportion

of the buildings himself and became a very wealthy man as a result. However, Hooke's later years were marred by scientific disputes, in particular with Isaac Newton (see pp.76–81), whom he accused of stealing his work without acknowledgment, and he chose to pursue a largely reclusive life.

THE **DRAWING** OF **THE FLEA** IN *MICROGRAPHIA* **WAS 18 IN (46 CM) WIDE**

HELD POSITION AS THE **ROYAL SOCIETY'S CURATOR** OF **EXPERIMENTS** FOR MORE THAN **40 YEARS**

Hooke's compound microscope had a biconvex lens, two extra lenses, and a focusing screw. Hooke also devised lighting for his specimens which, together with the microscope's precision, enabled him to reveal the structure of organisms as small as this flea.

ANTONIE VAN **LEEUWENHOEK**

Dutch textile merchant and amateur scientist Leeuwenhoek was a pioneer of microscopy. He built his own short-focal microscopes and was the first person to observe bacteria.

Leeuwenhoek (1632–1723) designed his own microscopes with a single, high-quality lens that he ground himself. Whereas Hooke had examined known organisms, Leeuwenhoek explored the invisible world. In 1674, he observed tiny creatures he named "animalcules" (protozoa) in specimens including wood, semen, and blood, and in 1676 noted single-celled organisms in water—the first recorded view of bacteria. He conveyed his findings via letter to the Royal Society and became a member in 1680. Leeuwenhoek made more than 500 microscopes, with magnification ranging from 70 times to 266 times, but his lens-making technique remained a lifelong secret.

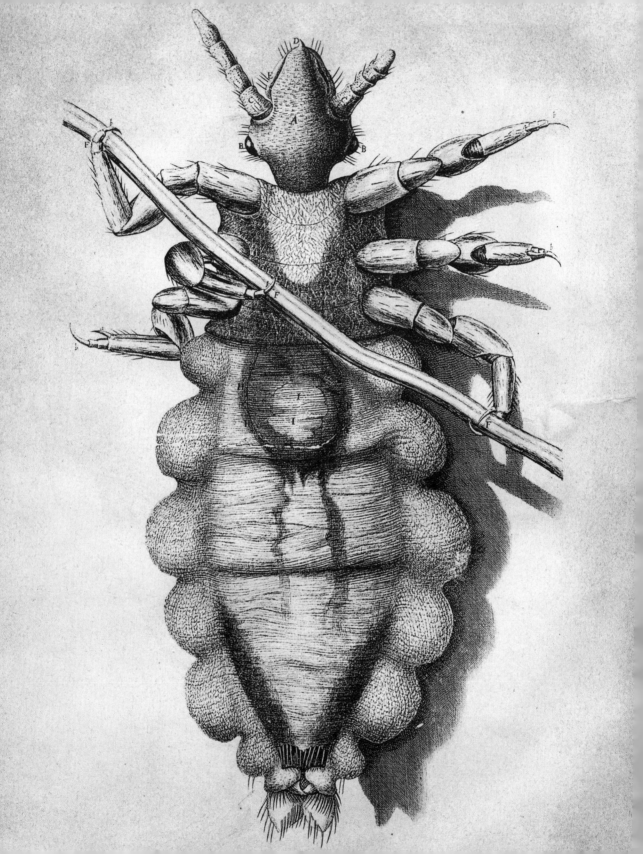

"BY THE MEANS OF TELESCOPES, THERE IS NOTHING SO FAR DISTANT BUT MAY BE REPRESENTED TO OUR VIEW ... BY THE HELP OF MICROSCOPES, THERE IS NOTHING SO SMALL, AS TO ESCAPE OUR INQUIRY; HENCE THERE IS A NEW VISIBLE WORLD DISCOVERED TO THE UNDERSTANDING."

Robert Hooke
Micrographia, 1665

◄ *Engraving of a human louse* as observed
by Robert Hooke and published in Micrographia.

ISAAC NEWTON

British physicist and mathematician Isaac Newton's laws of motion and law of universal gravitation fundamentally changed the way that humans understood the world. His groundbreaking work revolutionized scientific thought, making him one of the greatest scientists to have ever lived.

Born in Lincolnshire on Christmas Day in 1642, Isaac Newton was brought up there by his grandmother, and suffered from loneliness and ill health. He later attended Cambridge University, but in 1665, it closed due to the Plague, and Newton returned to his home county to avoid contagion. It was here that, reputedly, his transformative theory of gravity first began to evolve, and from an inauspicious start, Newton would rise to become one of the world's most influential scientists.

Planetary motion

Newton's great theory was sparked by his observation of an apple falling from a tree in his grandmother's orchard. The questions this generated—as to why the apple fell downward and what was controlling that movement—gave rise to his three foundational laws of motion and his law of universal gravitation, which together became the backbone of scientific understanding of how the world works. In the 17th century, the accepted view of the Universe was that it was heliocentric, meaning

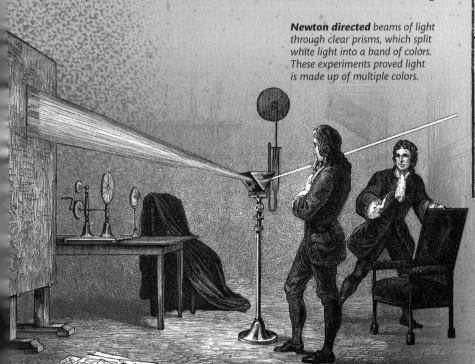

Newton directed beams of light through clear prisms, which split white light into a band of colors. These experiments proved light is made up of multiple colors.

MILESTONES

GRAVITY INSPIRATION
Observes an apple falling from a tree and begins to formulate his law of universal gravitation in 1666.

REFLECTOR TELESCOPE
Revolutionizes the design of telescopes by using mirrors to reflect light and clarify image in 1668.

KEY PUBLICATION
Publishes *Principia* in 1687, his seminal work outlining three laws of motion and the law of gravity.

SCIENTIFIC LEADER
Accepts the position of President of the Royal Society in 1703; holds the position for 24 years.

SCIENCE OF LIGHT
Discovers the composition of white light; publishes the results in his 1704 text entitled *Optiks*.

that the planets revolved around the Sun—but no explanation existed for *why* the planets moved in orbit. Newton solved this problem. From his observations in the orchard, he concluded that there must be an invisible force pulling the apple (and therefore all objects with mass) to the Earth. He further theorized that the force affecting movement of the apple must also affect celestial bodies, such as the Sun, Moon, and planets. Newton became the first scientist to understand that it was gravity that controlled the movement of the planets and held them in orbit.

Newtonian Laws

In 1687, Newton published his findings in his landmark text *Philosophiae Naturalis Principia Mathematica*, one of the most important books in scientific history. In it, he describes his three laws of motion, which act as a mathematical explanation of how speed and mass affect the movement of an object. The laws state that a static object will move if force is exerted on it; that acceleration depends on the force exerted and the mass of the object; and that all forces act in pairs that are equal and opposite. His law of gravity states that gravity is a universal force and that all objects (including celestial ones) exert a gravitational pull on each other. This represented one of the most significant scientific breakthroughs of all time.

Other fields

Newton extensively studied the nature of light and made great advances, which he published in his 1704 text *Optiks*. Through experiments with clear glass prisms, he revealed that white light is composed of all colors of the spectrum. He also pioneered the use of light through telescopes by designing the first reflecting telescope in 1668, which used mirrors (instead of lenses) to reflect light and create a sharper image than had previously been possible.

Beyond science, Newton applied himself to the study of alchemy and reinterpretations of the Bible.

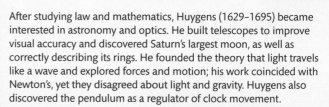

CHRISTIAAN **HUYGENS**

An influential figure in the 17th-century scientific revolution, Christiaan Huygens was a Dutch scientist and mathematician whose discoveries in physics and astronomy left a lasting legacy.

After studying law and mathematics, Huygens (1629–1695) became interested in astronomy and optics. He built telescopes to improve visual accuracy and discovered Saturn's largest moon, as well as correctly describing its rings. He founded the theory that light travels like a wave and explored forces and motion; his work coincided with Newton's, yet they disagreed about light and gravity. Huygens also discovered the pendulum as a regulator of clock movement.

Although solitary and difficult by nature, he held many public positions, such as President of the Royal Society, Master of the Royal Mint, and Member of Parliament. He is regarded as a scientist of unparalleled genius, whose revolutionary ideas laid down the foundations of modern scientific thought.

"I now **demonstrate** the **frame** of the **System** of the **World.**"

Isaac Newton, 1687

DEVELOPED THE **FOUNDATIONS** OF **CALCULUS** AGED **24**

WROTE **10 MILLION** WORDS OF TEXT

Newton explained the motion of both falling apples and planets with his law of universal gravitation. He determined that the force that pulled objects to the center of the Earth existed between all objects.

"WHY SHOULD THAT APPLE ALWAYS DESCEND PERPENDICULARLY TO THE GROUND? ... THERE MUST BE A DRAWING POWER IN MATTER, AND THE SUM OF THE DRAWING POWER IN THE MATTER OF THE EARTH MUST BE IN THE EARTH'S CENTER."

Isaac Newton
Related to William Stukeley, author of *Memoirs of Sir Isaac Newton*, 1726

*This illustration from Arthur Mee's **Popular Science** reimagines Newton observing the apple fall in 1666.* ▶

BENJAMIN

FRANKLIN

Self-educated and of humble origins, Benjamin Franklin was an inventor, scientist, printer, and politician. Through his scientific experiments, he made significant advances in our understanding of electricity and produced several key inventions. He is also revered as one of the Founding Fathers of the United States.

The son of a Bostonian soap- and candle-maker, Benjamin Franklin was 10 years old when his formal education ended, but his naturally enterprising spirit saw him thrive in multiple occupations, from writer and politician to inventor and scientist. His professional life began at the age of 12, when he was apprenticed to his older brother, who was a printer.

Franklin had an innate aptitude for learning and threw himself into the world of books, reading avidly and improving his writing skills. He became an adept printer and satirical columnist before relocating to Philadelphia, where in 1728, he set up as a printer himself. The financial success he enjoyed was so great that by 1748, he retired from business and applied himself to personal pursuits, including philanthropic projects and politics, but particularly science. He went on to make significant contributions to the field of electricity, for which he became internationally famous.

Experimenting with electricity

In the 1700s, electricity was little understood. It was mainly seen in tricks performed by showmen, and many believed that it was supernatural. In 1746, Franklin witnessed its effects at a lecture given by the scientist Archibald Spencer and was immediately drawn to discover the science behind it. To that end, he bought Spencer's equipment and conducted his own experiments. His findings led him to conclude that the prevailing theory of electricity—that it was

MILESTONES
PRINTING BUSINESS Sets up as a printer in Philadelphia and becomes the official printer of Pennsylvania in 1730.
BECOMES SCIENTIST Accrues sufficient wealth by 1748 to retire from business and focus on his scientific experiments.
ELECTRICITY STUDY Publishes a text on his early experiments in electricity in 1751, and proposes his kite experiment in 1752.
OCEAN MAPPING Charts and names the Gulf Stream between 1764 and 1765 so ships can avoid it to speed their passage.

While experimenting with various kinds of equipment, Franklin noticed that pointed electrodes were better at conducting electricity through air than spherical ones.

"Do not squander Time; for that's the Stuff Life is made of."

Benjamin Franklin, 1746

a "fluid" that came in two kinds, one that attracted and one that repelled—was incorrect. Instead, he argued that there was only one kind of electricity, which had either a positive or a negative charge. He further proposed that electricity could be neither created nor destroyed, but only transferred from one object to another, while its total amount of charge remained the same. This idea later became known as the law of the conservation of charge.

Natural electricity

Franklin also saw a correlation between thunder and lightning and the bangs and sparks generated by his experiments. He concluded that lightning must be natural electricity. Having previously noted that electricity was attracted to needle points, he proposed an experiment in which an iron rod was mounted on a large insulated box to attract a lightning bolt. When the rod was struck by lightning, it produced

In Franklin's famous experiment, electrical charge was attracted down the kite's wet string as it formed a natural conductor. Raising his hand to an attached key, he noticed a spark. He inferred that the charge came from electricity within the cloud.

> "Electricity ... **may** help to keep a **vain Man** humble."

Benjamin Franklin, 1747

a spark inside the box. This experiment was later repeated by several scientists, albeit with one fatality.

With the help of his son, Franklin also conducted an experiment with a kite in a thunderstorm. The kite was attached to some wet string, which had a key tied to its lower end. As the kite entered the thunderclouds, the electrical charge in the clouds was drawn down the string, causing the key to emit sparks, demonstrating that the clouds were electrified. Having shown that lightning and electrical charge were the same, Franklin published his theories and experiments in 1752, to global acclaim. His work also led to his invention of the lightning conductor, which, when attached to buildings, conducts lightning safely to Earth.

Inventions and beyond

Franklin devised several other inventions, including a heat-efficient stove, bifocal glasses, and a "glass armonica" (a musical instrument that uses glass bowls to produce notes). He also charted the current that flows eastward across the North Atlantic Ocean, naming it the Gulf Stream. A high-ranking diplomat heavily involved in government politics, Franklin also helped in drafting the Declaration of Independence, making him one of America's Founding Fathers.

HIS RESEARCH LED TO THE INVENTION OF THE BATTERY

REFUSED TO PATENT HIS **INVENTIONS** AND GAVE THEM TO THE **WORLD** FOR **FREE**

In the mid-18th century, Swedish botanist and physician Carl Linnaeus revolutionized the way scientists categorized and named living organisms by devising a uniform, hierarchical taxonomic system, which is still in use today.

Linnaeus's favorite plant, Linnaea borealis (twinflower), *was named after him. Its two-part name indicates its genus and species, a division that Linnaeus himself devised.*

Carl Linnaeus was born in a poor area of southern Sweden, where his father was a curate and amateur botanist. While he showed little interest in either his formal education or an ecclesiastical career, his enduring passion and fascination for plants was evident from a young age. After completing his schooling, he studied botany and medicine at Uppsala University and, due to his extensive knowledge of botany, began lecturing at the university 2 years later. In 1732, Linnaeus received funding for a research expedition to Lapland, where he intended to record and gather information on natural resources and discover new species of plants and animals. The results of this expedition were to inspire him to produce a groundbreaking manuscript that would revolutionize the natural sciences.

Categorizing life

The inconsistent and overcomplicated systems of classification that existed at the time frustrated Linnaeus in his botanical studies. Attempts to classify life forms had begun with Aristotle, and while multiple disordered adaptations had appeared over the centuries, by the 18th century, there was still no unified system. During his expeditions in northern Sweden, Linnaeus formulated a method of categorizing different species that would simplify and transform the process.
In 1735, he published the *Systema Naturae*,

LINNAEUS

a landmark work that provided a new classification system for the natural world. Beginning as an 11-page pamphlet, the *Systema Naturae* passed through 12 editions during Linnaeus's lifetime to become an extensive, multivolume text. In it, he grouped all living organisms according to their shared physical characteristics, then organized these groups into a hierarchy of increasing exclusiveness.

At the top of the hierarchy were the largest groups: three wide-reaching "kingdoms," which divided nature into animals, plants, and stones. From there, it extended down through different subdivisions as far as "species," the final group containing just one organism. This formalized structure—recognized today as domain, kingdom, phylum, class, order, genus, family, and species—enabled scientists to compare and identify millions of different organisms and rationalize the natural world.

Arguing that all living organisms reproduced sexually, Linnaeus grouped the plants on this basis. He also began a new phase in biological thought as he classified humans alongside other life forms and defined them as animals for the first time.

Labeling life
In several other seminal botanical texts that Linnaeus published, including *Genera Plantarum* and *Species Plantarum*, he completely overhauled and reorganized how different organisms were named. This was arguably his most significant contribution to the field of botany. Previous systems had included complex descriptive phrases,

as well as common names, which Linnaeus stripped back and streamlined. He provided each organism with a two-part Latin name (or binomial), which included the genus and the species—for example, *Homo* (genus) *sapiens* (species). This universal naming method—initially applied to plants, but later extended to animals in the 10th edition of *Systema Naturae*—was succinct, descriptive, and specific, and it established Latin as the official language of taxonomy.

Taxonomic legacy
Linnaeus's reorganization of the natural world marked a turning point in biological science. Although he devised it to reflect what he regarded as God's divine plan of creation, today it forms the foundation of modern taxonomy—the

In his quest to catalog the various species of the world, Linnaeus recorded 6,000 species of plants and 4,000 species of animals.

science of classifying, identifying, and naming organisms. Although Linnaeus's hierarchical system was refined over time to include additional groupings and subdivisions, the basic Linnaean system remains fundamentally the same today. For this reason, Linnaeus is considered the father of taxonomy.

> "Nature does not **proceed** in leaps and **bounds.**"

Carl Linnaeus, 1751

NAMED MORE THAN **10,000** LIFE FORMS

EXTENDED HIS BOOK, *SYSTEMA NATURAE*, FROM **11** TO **3,000** PAGES

HIS **NAMING SYSTEM** STILL USED **250** YEARS LATER

GASPARD **BAUCHIN**

Linnaeus was greatly influenced by Swiss botanist and anatomist Gaspard Bauchin.

Bauchin (1560–1624) gained degrees in medicine, anatomy, and botany, becoming a professor at the University of Basel. He was a practicing doctor and produced influential texts in both anatomy and botany. *Illustrated Expositions of Plants*, published in 1623, was a pivotal early catalog of plant life that listed 6,000 different species. Bauchin named each plant according to its genus and species, a practice later adopted by Linnaeus.

89

Doct: LINNÆI
METHODUS plantarum SEXUALIS
in SYSTEMATE NATURÆ
descripta

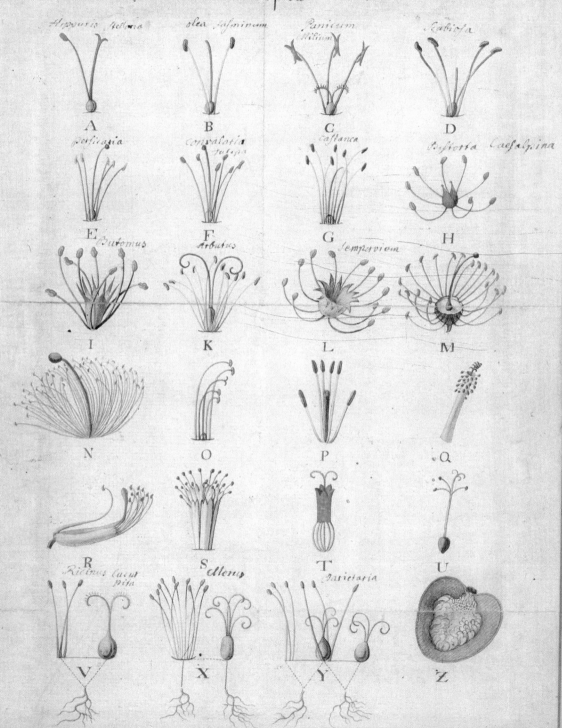

"I DEMAND OF YOU, AND THE WHOLE WORLD, THAT YOU SHOW ME A GENERIC CHARACTER— ONE THAT IS ACCORDING TO GENERALLY ACCEPTED PRINCIPLES OF CLASSIFICATION, BY WHICH TO DISTINGUISH BETWEEN MAN AND APE. I MYSELF MOST ASSUREDLY KNOW OF NONE ..."

Carl Linnaeus
In a letter written to J. G. Gmelin, 1747

◄ *In* **Systema Naturae (1735)**, *Linnaeus classified plants based on the number and arrangement of stamens in a plant's flower.*

JAMES HUTTON

Geologist James Hutton was the first to consider the age of the Earth scientifically. His fieldwork in the Scottish landscape revealed evidence to demonstrate that the Earth was far older than had been thought. His work laid the foundations for the modern discipline of geology.

James Hutton was born in Edinburgh to the family of a British merchant in 1726. After showing interest in mathematics and chemistry, he became a physician's assistant at 18 and went on to study medicine in Edinburgh, Paris, and the Netherlands. Rather than becoming a doctor, Hutton then set up a chemical plant and modernized his family's farms.

Hutton's fieldwork, allied with his agricultural and chemical knowledge, led him to lay the founding principles of geology. "Unconformity" was his term for the junction between different rock strata (layers) that had been laid down at different times. Unconformities can be seen in cliffs and outcrops as the rock layers above them differ from those below. Hutton also realized that rocks were formed under intense subterranean heat and pressure in a "uniformitarian" process, meaning that it occurs at the same rate today as it has always done. This made it possible to estimate the age of rock formations, and therefore the Earth, leading him to conclude that the Earth was far older than the 6,000 years proposed by biblical scholars.

MILESTONES

IN THE FIELD
Tours the north of Scotland in 1764 with George Clerk-Maxwell in search of geological features.

DIGGING DEEP
In 1767, studies rock strata revealed in excavations for the Forth and Clyde canal, Scotland.

UNIFORM THINKING
Publishes theory of uniformitarianism in a paper for the Royal Society of Edinburgh in 1785.

THEORY OF THE EARTH
Claims in 1788 that much of the land was once under the sea, and layers of rock are distorted over time.

Hutton used the term "unconformity" for a gap in the rock sequence at the point where the vertical layers meet the horizontal ones, as seen in his 1787 sketch of a rock face from Jedburgh, Scotland.

Wealthy British aristocrat Henry Cavendish was an outstanding theoretical chemist and physicist whose meticulous research established the true nature of air and water and the properties of hydrogen and carbon dioxide. He also made the first calculation of the density, and hence the weight, of the Earth.

Henry Cavendish studied natural philosophy (science) at Cambridge University and then settled in London, where he played an active role in scientific organizations, particularly the Royal Society, which he joined in 1760. He started conducting experiments in 1764, exploring many areas of science—including electricity, chemistry, and physics—although he published only a few of his findings during his 50-year career.

Isolating hydrogen

In the 18th century, chemists started to investigate ways of generating and separating various gases, although they thought these substances were different types of air rather than unique chemicals. Cavendish's first publication, in 1766, was a combination of three chemistry papers on "factitious airs," or gases, created in a laboratory. In one of the rigorous investigations that characterized his work, he had dissolved zinc, iron, and tin in acids and found this produced a gas that he called "inflammable air." It was ultimately given the chemical name hydrogen. The most abundant element

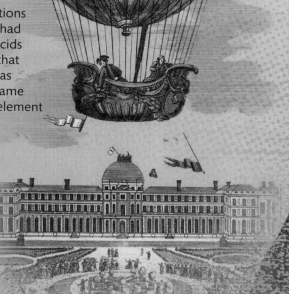

In 1783, after studying the gas chemistry research of Cavendish and others, Parisian Jacques Charles developed the world's first hydrogen-filled balloon, starting a ballooning craze.

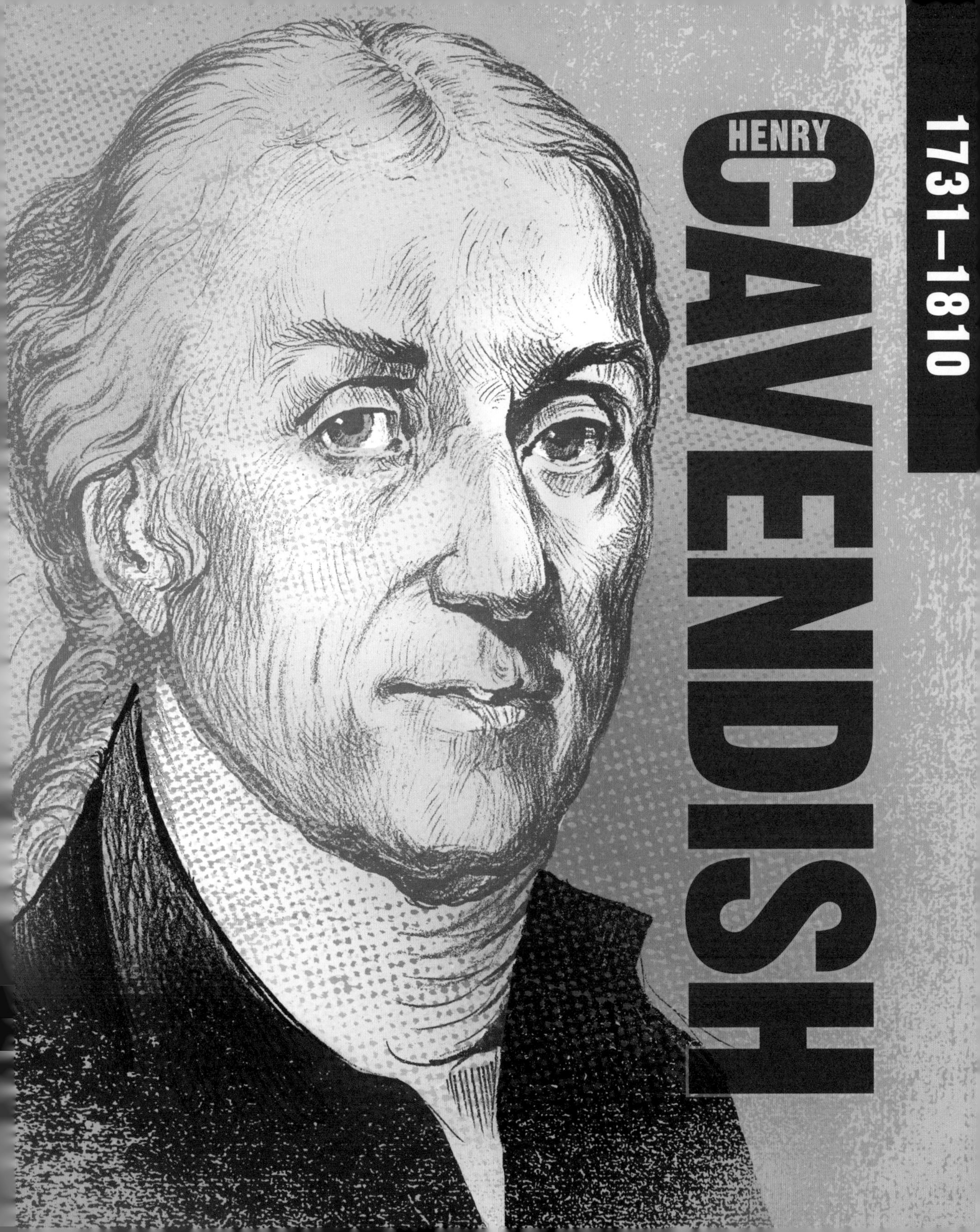
HENRY
CAVENDISH
1731–1810

in the Universe, hydrogen had been detected and burned by earlier scientists, but Cavendish was the first to recognize it as a distinct substance. He also isolated carbon dioxide, which was then known as "fixed air." He compared hydrogen and carbon dioxide's chemical and physical properties with those of ordinary air. To determine hydrogen's density, he filled a pig's bladder with it and, weighing that, calculated the gas to be 11 times lighter than air (slightly less than the modern accepted figure). He found carbon dioxide to be 1.57 times heavier than air and dissolvable in water.

Measuring water and air

Cavendish then turned to exploring the composition of ordinary, or atmospheric, air. He analyzed air on 60 occasions and from different places, even using a balloon to collect samples from the atmosphere. He expressed his results in terms of Johann Becher's phlogiston theory, which stated that all combustible matter contained an odorless and colorless substance called "phlogiston." When matter was burned, the phlogiston was released, after which the remaining matter, now ash or residue, was considered to be "dephlogisticated."

Correctly reporting that air is four parts "phlogisticated air" (nitrogen) to one part "dephlogisticated air" (oxygen), Cavendish also found a scant amount of a gas that he could not identify. A century later, this was identified as argon.

Having observed that burning his inflammable air (hydrogen) also produced water, in 1784, Cavendish published a paper that revealed water's chemical formula—two parts inflammable air (hydrogen) to one part dephlogisticated air (oxygen). He had proved that water is

not a chemical element—a single, distinct substance—as had been thought for millennia, but a compound of two gases.

Cavendish experiment

Between 1797 and 1798, Cavendish devised an ingenious experiment to determine the density of the Earth—an elusive value sought by both scientists and explorers. His apparatus for "weighing the world" was a wooden rod suspended from a wire, with a small lead sphere on either end and two much larger lead spheres fixed in place close to the smaller spheres. Since all objects with mass exert a gravitational force on each other, the large spheres attracted the smaller ones, causing a slight torque (twist) on the suspended

Cavendish's experiment, which ultimately enabled him to calculate the density of the Earth, also proved that Newton's law of gravitation works with objects much smaller than the planets.

rod. Cavendish viewed the oscillations on the rod through a telescope; because the force of gravity is very weak, he was aware that the subtlest disturbance in the two sides of his equipment, even a shift in temperature, might skew the results, so he had placed it in a dark, sealed room. After a year of painstaking observation, Cavendish determined the gravitational force between the pairs of spheres. Knowing the density of the spheres, and knowing the gravitational pull of the Earth, he was then able to produce a value for the density of the Earth, which turned out to be more than five times that of water.

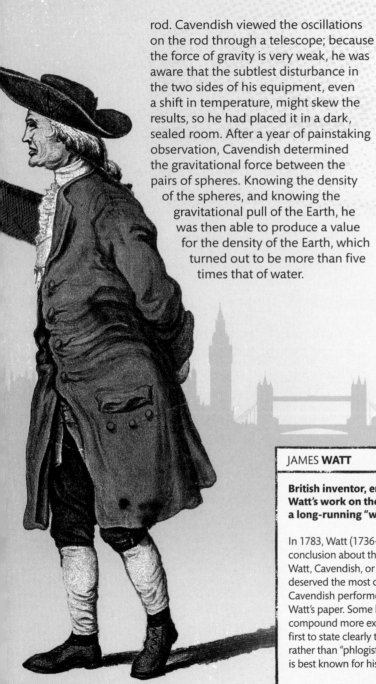

CONDUCTED MORE THAN 400 INVESTIGATIONS INTO THE NATURE OF AIR

PROVED THAT THE DENSITY OF EARTH WAS 5 TIMES THAT OF WATER

$8,230 OF HIS LEGACY WAS USED TO BUILD THE CAVENDISH LABORATORY AT CAMBRIDGE

JAMES WATT

British inventor, engineer, and chemist James Watt's work on the composition of water led to a long-running "water controversy."

In 1783, Watt (1736–1819) and Cavendish reached almost the same conclusion about the chemical make-up of water. But whether Watt, Cavendish, or French chemist Antoine Lavoisier (see pp.100–103) deserved the most credit for this discovery became the subject of debate. Cavendish performed his experiments first but published them after Watt's paper. Some historians say that Watt identified water as a chemical compound more explicitly than Cavendish. However, Lavoisier was the first to state clearly that water is composed of hydrogen and oxygen rather than "phlogiston" (the term used by Watt and Cavendish). Watt is best known for his patented improvements to steam engines.

WILLIAM
HERSCHEL

A German-born astronomer living in England, William Herschel discovered Uranus, the first new planet to be found since the invention of the telescope. He was a skillful engineer who designed and built the world's largest telescope at the time.

William Herschel's interest in astronomy began after he learned how to make telescopes in order to study the night sky. Determined to look beyond the closest planets, he built a powerful telescope through which he could study the farthest celestial bodies. Herschel made detailed observations of the night sky, methodically dividing it into quarters and recording his sightings systematically. In 1781, he noted a small disk, which was confirmed as the seventh planet in the solar system. Named Uranus, its discovery led to the eventual sighting of Neptune in 1846.

Further discoveries

Astronomy in the 19th century profited from Herschel's considerable contributions. From his observations, he argued that nebulae (clouds of gas and dust) are composed of stars, not fluid, and he went on to catalog 2,500 nebulae and star clusters and 848 double stars. He discovered the two moons of Uranus, as well as two new moons of Saturn, and was the first to discover infrared rays after making studies of the Sun. Herschel's exceptional skill as a telescope builder, combined with his meticulous observations of the night sky, made him one of the greatest visual astronomers of all time.

MILESTONES

DEVELOPS INTEREST
Moves to England in 1757; becomes interested in astronomy and observes the night sky.

FINDS NEW PLANET
In 1781, discovers Uranus and is elected a Fellow of the Royal Society; now a professional astronomer.

OBSERVES NEBULAE
Publishes first of three papers on nonstellar objects in 1786; redefines understanding of nebulae.

BUILDS TELESCOPE
In 1789, constructs an enormous, 40-ft (12-m) telescope with a 47-in (122-cm) mirror.

INFRARED RAYS
Discovers, in 1800, that the Sun emits invisible "calorific rays," now known as infrared rays.

> **"I have looked further into space than ever human being did before me."**
>
> **William Herschel**, 1781

Herschel captured the public imagination with his telescope, built in 1789, but it worked poorly and he considered it a failure.

ANTOINE-LAURENT
DE LAVOISIER

Antoine-Laurent de Lavoisier's meticulous experiments sparked a revolution in science, and his findings had a vital impact on chemistry. His work on combustion and respiration allowed him to determine his law of conservation of mass, and to discover that air was a mixture of gases, including one that he called oxygen.

Born in Paris, Antoine-Laurent de Lavoisier inherited a large fortune at age 5, when his mother died. He qualified as a lawyer but was more interested in science, which he pursued whenever he was able. In 1768, he bought a share in the Ferme Générale, which collected taxes on behalf of the French government, and in 1771, he married the co-owner's 13-year-old daughter, Marie-Anne. She became an invaluable partner in his scientific work, translating chemistry papers into French and assisting him with his experiments.

Investigating combustion

When he became commissioner of the Royal Gunpowder Administration in 1775, Lavoisier gained access to a laboratory in which he could conduct experiments on combustion. At that time, an incorrect idea called phlogiston theory held that combustible matter such as wood contained "phlogiston," an invisible substance that was released on burning, which explained why substances apparently lost mass when burned. However, Lavoisier found that when he burned sulfur and phosphorus, they acquired mass, suggesting that they were combining with air, and the results could not be explained in terms of phlogiston.

A year earlier, in 1774, Lavoisier had learned from English natural philosopher Joseph Priestley how he had heated mercury calx (mercury oxide) and collected what he called "pure" or "dephlogisticated" air, which burned much more

MILESTONES

ELEMENT FORMS
Proves elements can have different forms by burning a diamond to produce carbon gas in 1772.

INVESTIGATES AIR
Hears from Joseph Priestley about "dephlogisticated air" in 1774, and replicates his findings.

ANALYZES WATER
In 1783, he reveals that water is a compound of two gases and coins the word "hydrogen."

CHEMISTRY TEXTBOOK
Publishes *Elementary Treatise on Chemistry* in 1789—it contains the first table of elements.

In an experiment, Lavoisier used a giant magnifying glass to focus the Sun's rays on a diamond in a sealed glass jar. The diamond burned away, but the jar's mass stayed the same, proving his theory of the conservation of mass.

> # "We must **trust to nothing but facts:** these are **presented** to us **by nature** and **cannot deceive."**

Antoine-Laurent de Lavoisier, 1790

vigorously than ordinary air. Back in his laboratory, Lavoisier replicated Priestley's experiment. He heated mercury calx and measured how much "pure air" (or "oxygen," as he later named it) was released as the calx broke down into mercury. He also heated the mercury and measured how much oxygen was taken up as it changed into mercury calx again. The amount of oxygen taken up and released matched.

When he heated other substances in sealed containers, Lavoisier found that the mass a metal gained when heated was exactly equal to the oxygen lost from the air. As a result of these experiments Lavoisier started to form his law of conservation of mass, which stated that matter could change in form but could not be created or destroyed.

Air, water, and elements

Lavoisier observed that a lit candle in a sealed jar would extinguish after a time and, from his various experiments on combustion, he knew that oxygen had allowed the candle to burn. However, he was also aware that some gas remained in the jar and deduced that air must contain several gases. He called the other gas "azote," meaning "without life." (It was later called "nitrogen.") Lavoisier also

investigated respiration, which he thought converted oxygen into energy in the form of heat and an azotic gas (carbon dioxide) in the same way that a burning candle does. Lavoisier placed a guinea pig in a container that was surrounded by ice and measured the heat and carbon dioxide that the animal produced while 2 pounds of ice melted. He then compared these results with the amount of heat produced by burning carbon until the same amount of carbon dioxide was given off as the guinea pig had produced. In this way, Lavoisier showed that respiration (breathing) was also a form of combustion.

By comparing the weights of substances before and after chemical reactions, Lavoisier determined the law of conservation of mass: matter only changes forms; it cannot be created or destroyed.

Lavoisier also repeated an experiment carried out by English scientist Henry Cavendish (see pp.94–97), who created a gas he called "inflammable air" when he poured acid onto metals. When Cavendish applied heat to the gas, water was formed. In his experiment, Lavoisier found that the heat acted as a catalyst and made the "inflammable air" (which he renamed "hydrogen") react more quickly with oxygen in the air to create water. Lavoisier was the first scientist to definitively state that water is made up of hydrogen and oxygen rather than "phlogiston."

JOSEPH **PRIESTLEY**

English clergyman Joseph Priestley discovered 10 gases, including "dephlogisticated air," which Lavoisier named "oxygen."

Living near a brewery, Priestley (1733–1804) saw something he called "heavy air" (carbon dioxide) bubbling up from the grain and noted it could dissolve in water. He later added mint to a container of carbon dioxide and found it "refreshed" the air sufficiently to support a lighted candle—he had discovered plants emit oxygen. In 1774, Priestley told Lavoisier about his experiments with mercury and "dephlogisticated air" (oxygen). Although his work inspired Lavoisier, Priestley rejected Lavoisier's theories and continued to believe substances contained "phlogiston."

HEATED
MERCURY OXIDE FOR **12 DAYS** TO DETERMINE THAT **AIR** CONTAINS **OXYGEN**

HIS **TABLE OF ELEMENTS** INCLUDED **OXYGEN** AND **HYDROGEN**, BUT ALSO **LIGHT** AND **HEAT**

Italian aristocrat Alessandro Volta was a meticulous investigator whose experiments revealed that electricity can be generated chemically, as well as by friction. Chief among his many scientific contributions was the invention of the Voltaic pile, the world's first battery, which laid the path for the coming age of electric power.

Born in Como, Italy, Alessandro Volta was the only one of seven siblings not to become a priest, having met scientist and physician Giuliano Gattoni. At the age of 23, he wrote his first paper on electricity, and at 29, he became professor of physics at Como's Royal School.

Volta was quick to make scientific breakthroughs. In 1775, he improved the electrophorus, a resin disk that could be rubbed with cat fur to make a static charge—the only kind of electricity known at that time. But it was Luigi Galvani's 1786 finding that a dissected frog's legs could be made to twitch by electricity that sparked him into action. Rejecting this idea of "animal electricity," Volta theorized that contact between different metals created a chemical reaction, which produced electricity. He then tested different combinations of metals by placing them on his tongue, resulting in a mild electric shock. This led to his 1799 discovery that stacking disks of zinc, copper, and brine-soaked paper in sequence produced an electric current, which increased as more disks were added. His "Voltaic pile" was the first battery and the prime means of creating electricity until the discovery of electromagnetic generation in 1829.

MILESTONES

GAS DISCOVERY
Captures methane while collecting marshland air in 1776. Ignites the gas with a spark inside a glass jar.

ACADEMIC POST
In 1779, becomes professor of experimental physics at the University of Pavia, where he stays for 40 years.

MAKES ELECTRICITY
Generates electric current from a circuit of copper, zinc, and brine-soaked paper in 1791.

FIRST BATTERY
Creates the "Voltaic pile" in 1799, the first apparatus for generating and storing electricity.

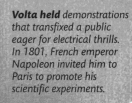

Volta held demonstrations that transfixed a public eager for electrical thrills. In 1801, French emperor Napoleon invited him to Paris to promote his scientific experiments.

While studying smallpox—a virus that killed 400,000 people a year in 18th-century Europe—British doctor Edward Jenner established the technique of vaccination, which has since saved billions of lives. His smallpox vaccine, the world's first successful vaccine, eventually led to the total eradication of the disease.

Edward Jenner was born in the county of Gloucestershire, the eighth of nine children. From the age of 5, following the death of his clergyman father, he was brought up by one of his older brothers. As a youth, he had his first brush with smallpox—the disease that would mark his future career—when he became ill after undergoing variolation. This form of inoculation involved inserting pus or a scab from a patient with mild smallpox into a scratch or the nostrils of a healthy person to boost their immunity. Although it was a popular treatment among the aristocracy in Europe, it carried a high risk of onward transmission or even death.

Medical training

Jenner was apprenticed to a local surgeon at the age of 14 and moved to London 8 years later to become a private pupil of John Hunter, a leading surgeon, anatomist, and biologist known for his scientific methods and groundbreaking experiments. A mentor and friend for

> "One of the greatest benefactors of mankind."
>
> **Napoleon Bonaparte**

Jenner demonstrated the link between cowpox and smallpox by conducting careful medical trials. His 1798 report detailed his treatment of 23 patients, including several children.

EDWARD JENNER

"**Future nations will know ... smallpox has existed and by you has been extirpated.**"

Thomas Jefferson, 1806

the rest of his life, Hunter instilled the young Jenner with the Enlightenment motto, "Don't think, try."

After 3 years of study in London, in 1773, Jenner qualified as a doctor and returned to his place of birth to set up a medical practice. Having grown up in an agricultural area, he was aware that catching cowpox—a relatively benign disease—somehow prevented smallpox. Milkmaids, for instance, rarely contracted smallpox if they had previously caught cowpox due to contact with their cows.

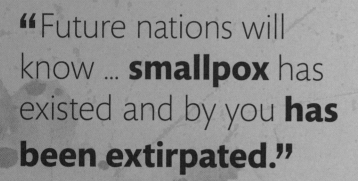

By giving patients cowpox to fight smallpox, Jenner discovered the principle of vaccination. When a body is invaded by a virus, like cowpox, it produces antibodies, which protect it against reinfection.

When Sarah Nelmes, a milkmaid, visited his practice with cowpox in 1796, Jenner decided to put his theory to the test. On May 14, he transferred blister fluid from Nelmes to both arms of James Phipps, the 8-year-old son of his gardener. Although Phipps became slightly ill over the next 9 days, by the tenth, he was well again. To find out if his experiment had worked, on July 1, Jenner deliberately infected Phipps with smallpox. To Jenner's relief, the boy not only survived, but further tests carried out throughout his long life proved he had developed immunity to smallpox.

Critical reception

Jenner published the full results of his study, which included vaccinations of 23 patients—including his own son—in 1798. Initially, his report was met with opposition and even ridicule among an uneducated public. Although the medical community had doubts about the ethics of his experimental method, Jenner's thorough, scientific approach was lauded. Over time, his procedures were accepted and improved and were used in America, the rest of Europe, and across the world. The rate of smallpox

mortality declined dramatically, and Jenner was recognized internationally for his work.

In 1853, three decades after his death, Jenner's findings finally convinced the British government to replace inoculation with compulsory vaccination. The smallpox vaccine has since saved several billion lives and triggered a global initiative in the 1970s that was the first, and still only, program to have eradicated an infectious disease.

BENJAMIN **JESTY**

Though his contribution only became clear following Jenner's more scientific achievements, Jesty was eventually recognized as one of the first successful smallpox vaccinators.

Jesty (1736–1816) was a farmer in Dorset, England, who had contracted cowpox due to his work with livestock. When a smallpox epidemic broke out in 1774, he took the unprecedented step of using a darning needle to transmit cowpox pus to scratches on the arms of his wife and two eldest sons. His sons suffered mild, local reactions, but his wife became gravely ill—although all three recovered fully. Initially mocked for his actions, Jesty was vindicated when his sons proved to be immune upon later exposure to the disease. His pioneering efforts came to light decades later, and in 1805, he was honored by the Original Vaccine Pock Institution in London.

SMALLPOX CASES HAD A **MORTALITY RATE** OF **80%** IN THE 18TH CENTURY

TESTED EARLY VACCINATION ON HIS **11-MONTH-** OLD SON

VACCINATIONS FOR SMALLPOX MADE **COMPULSORY** IN ENGLAND **30 YEARS** AFTER HIS DEATH

JOHN
DALTON

British schoolteacher and meteorologist John Dalton advanced the concept of atomic theory, which is a cornerstone of physics and chemistry. His discoveries about the nature of elements and atoms and the way they combine were essential to establishing chemistry as a modern science.

MILESTONES

WEATHER INSIGHTS
Begins a lifelong habit of keeping daily weather records, including pressure and wind speed, in 1787.

NATURE OF GASES
Observes that gases are affected by pressure and heat; in 1801, he forms his law of partial pressures.

DIFFERENCES IN ATOMS
Becomes the first person to calculate and make charts of relative atomic weights in 1803.

WIDELY HONORED
Receives Royal Society's medal in 1826. Elected to French and American Academies of Sciences.

Born into a Quaker family in Cumbria, England, John Dalton began work as a teacher aged 12. His first scientific paper described red-green color blindness; it is still sometimes called Daltonism. Both Dalton and his brother suffered from it, and he correctly believed it to be hereditary.

From a young age, Dalton kept records of rainfall, wind speeds, and air pressure. His meteorology textbook, published in 1793, was the first to correctly describe the hydrologic cycle (in which water evaporates from the oceans and falls as rain). Dalton also stated that water vapor did not combine chemically with air, as many thought, and that water's rate of evaporation depended on heat and wind speed.

Study of atmospheric gases
Dalton's interest in the weather led him to conduct experiments on the nature of atmospheric gases. In 1801, he gave a series of lectures in Manchester about his groundbreaking findings. These included the

Dalton's research into atmospheric gases led him to realize that the size and mass of atoms varies between elements. He made diagrams showing their structures and diameters.

LISTED
THE RELATIVE WEIGHTS OF
20
ELEMENTS

MADE OVER 200,000 OBSERVATIONS IN HIS WEATHER JOURNALS

observation that the pressure of a volume of gas in a container depends on its temperature, and that given sufficient pressure and low temperature, all gases become liquid.

All gases exert pressure on the walls of the container that holds them. Dalton observed that even if there were several gases in a container, each one exerted the same pressure as it would if it were alone in the container. The total pressure of the combined gases was equal to the sum of these partial pressures. This became known as Dalton's law of partial pressures and remains a fundamental principle in modern chemistry.

Atomic theory
The idea of atoms (tiny indivisible particles) dated back to the ancient Greeks, but it was assumed that atoms were identical in every element. However, Dalton had noticed from his experiments that oxygen absorbed less water vapor than nitrogen. He wrote of this: "I am nearly persuaded that the circumstance depends on the weight and number of the ultimate particles of the several gases." In other words, he wondered if oxygen and nitrogen had atoms of different weights.

If this was the case for oxygen and nitrogen, Dalton reasoned, it must be so for every element. He ran a series of experiments comparing the known elements with hydrogen (the lightest of the elements) and assigned them each a number that represented their relative atomic weights. He went on to create the first atomic table featuring all the elements known at the time, each with a pictorial symbol. The weights he calculated were not always accurate because he did not realize that atoms of the same element might combine. For example, he assumed that oxygen was a single atom rather than a molecule with two atoms. However, later scientists were able to calculate the weights more accurately, and Dalton's principles form the foundation of modern chemistry.

In his book *A New System of Chemical Philosophy* (1808), Dalton set out his atomic theory. He said atoms were indivisible and could not be created or destroyed; each element had atoms of a unique mass (weight); and atoms of different elements combined in chemical reactions to form compounds in simple whole-number ratios. The atomic theory was quickly accepted, and Dalton was elected to the Royal Society in 1822. A few years later, Jöns Jacob Berzelius refined Dalton's atomic table using letters rather than pictorial symbols, but Dalton preferred his own symbols and did not adopt the new system.

"Small particles called **atoms** exist and **compose all matter."**

John Dalton, 1808

Dalton included all 20 elements then known to science in his atomic chart. Some, such as potash, are now identified as molecular compounds; others, such as phosphorus, are never found in their elemental form on Earth.

ELEMENTS

HYDROGEN   1

NITROGEN   5

OXYGEN   7

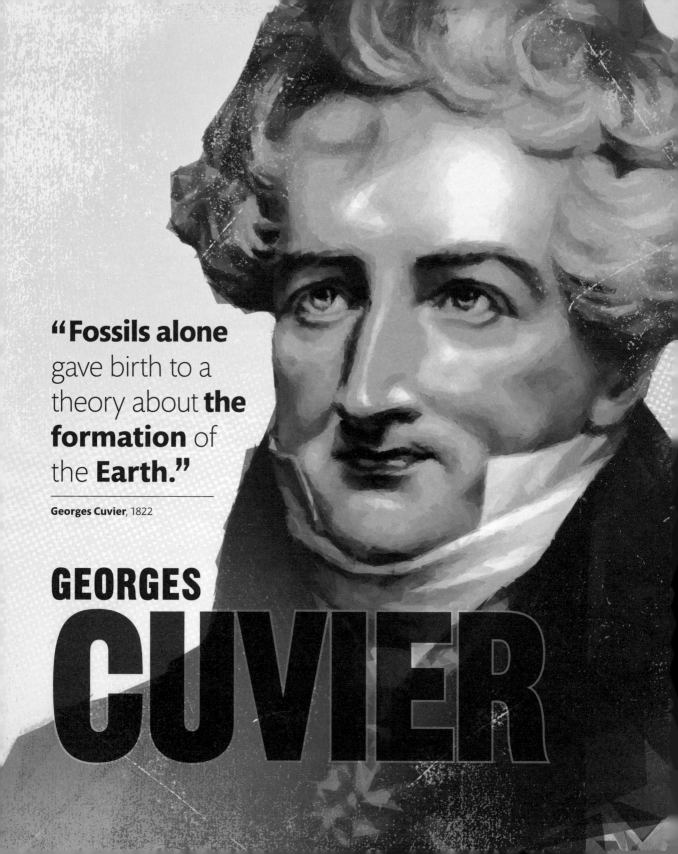

"**Fossils alone** gave birth to a theory about **the formation** of the **Earth**."

Georges Cuvier, 1822

GEORGES CUVIER

Regarded as the founder of paleontology, French zoologist and naturalist Georges Cuvier made huge advances in natural science during the late 18th and early 19th centuries. He established comparative anatomy as a scientific discipline and provided conclusive proof that animal species could become extinct.

MILESTONES

PROVES EXTINCTIONS
By comparing the bones of elephants with mammoths in 1796, he proves that extinctions occur.

AGE OF THE EARTH
In 1804, observes that the Earth must be far older than previously believed due to the age of fossils.

GROUPS ORGANISMS
Publishes his system of classifying organisms in 1817 and advances the field of animal taxonomy.

EARTH'S STRUCTURE
Proposes a theory in 1825 that past natural disasters affected the Earth's geological structure.

Georges Cuvier was born in Montbéliard, a town on the French-Swiss border which was then part of the German duchy of Württemberg, now in France. Educated at the prestigious Caroline Academy in Stüttgart, he studied comparative anatomy, a relatively new science examining similarities and differences between animal species. After writing detailed original studies of marine invertebrates, in 1795, Cuvier was offered a job at the National Museum of National History, Paris. There, he carried out the research that would establish him as one of the most influential figures of 19th-century natural sciences.

Theory of extinction
Using the museum's collection of animal specimens, Cuvier furthered his studies in comparative anatomy—in particular, by comparing the remains of living animals with fossils. The results he discovered were revelatory. Until the late 18th century, scientists had believed that fossils were the remains of extant (living) species, and that no animal species had ever

Many of Cuvier's publications were based on lectures he gave in Paris. In these, he captivated audiences with reconstructed skeletons of animals that no longer walked the Earth.

died out—Cuvier's research put an end to this belief. By comparing the fossilized remains of mammoths with those of modern elephants, he demonstrated for the first time that not only were the African and Asian elephants distinct species, but that the anatomy of a mammoth was completely different from either of theirs. From this, he concluded that the mammoth was in fact an extinct species, and went on to identify other extinct animals, including the giant ground sloth and the American mastodon. By establishing extinction as a verifiable fact, Cuvier launched the discipline of vertebrate paleontology.

Animal reconstruction

Renowned for his ability to reconstruct animals from their remains, Cuvier developed a theory he called the "correlation of parts," which stated that the structure of an organ in an animal's body was related to its function, as were all the other organs. For example, all hoofed mammals were herbivores and therefore must have teeth and a digestive system that is appropriate for eating plants. Using this principle, Cuvier was able to reconstruct complete skeletons of extinct organisms from just their isolated bones. He was also the first to identify the fossil of a flying reptile in 1800, which he named *pterodactyl*.

> ## "The component parts must be such as to render possible the whole living being."

Georges Cuvier, 1817

To explain extinction, Cuvier proposed the theory of catastrophism—that a series of sudden, catastrophic events in the Earth's history, such as flooding, had wiped out certain species. He used his own observations of changes in rock strata to support this. Although his theory was later disproved, in exhuming fossils from rock strata, Cuvier revealed a significant fact: the deeper the rock stratum, the older the fossils it contained. Cuvier's work indicated that the deepest fossils were thousands of centuries old and increased the age of the Earth far beyond the previously accepted age of 6,000 years.

Anatomical classification

Cuvier publicly supported catastrophism over the new theory of evolution and used his seminal work

PRIOR TO **CUVIER'S** **STUDIES** OF **FOSSILS**, SCIENTISTS AGED THE EARTH AT **6,000 YEARS**

HIS NAME IS ONE OF **72** INSCRIBED ON THE **EIFFEL** TOWER

Cuvier studied the skeletal remains of elephants and woolly mammoths, revealing key differences in tooth structure that identified them as distinct species and showed that the mammoth was extinct.

The Animal Kingdom as proof against it. This text classified animals according to anatomy, dividing them into four groups: vertebrates, mollusks, articulates (such as arthropods), and radiates (those with radial symmetry, like starfish).

Cuvier insisted that animals did not change over time, refuting the possibility of evolution. While this view was later disproved, Cuvier's system radically advanced the understanding of animal anatomy. His work forms the foundations of paleontology and comparative anatomy.

LOUIS **AGASSIZ**

A Swiss-born naturalist and geologist, Louis Agassiz made significant scientific advances through his work on glaciers and fossilized fish.

The extensive work of Louis Agassiz (1807–1873) into fossilized fish during the 1830s proved highly influential in the study of extinct life forms. His later geological studies of glaciers led him to conclude that much of the world had been covered in a vast ice sheet until relatively recently. A supporter of Cuvier's catastrophism theory, he cited ice ages as types of catastrophe that had affected the Earth. While these ice-age theories are supported by modern geology, Agassiz, like Cuvier, rejected the idea of evolution.

DIRECTORY

With the Enlightenment came a focus on reason, skepticism, and liberty that encouraged the systematic exploration of theories, and science became a distinct academic discipline. Increasingly sophisticated equipment, methods, and proofs emerged, as did some of the first modern scientific institutions.

MARCELLO MALPIGHI
1628–1694

In about 1661, while studying frog lungs under a microscope, Italian physician and biologist Marcello Malpighi spotted capillaries—tiny blood vessels that link arteries and veins. In the course of his work, he developed new methods for studying microscopic things, such as illuminating specimens or staining them so they could be seen more easily. His pioneering work in microanatomy led to him becoming an honorary member of the Royal Society in 1669, but Malpighi continued to practice medicine throughout his career. He became the pope's physician in 1691.

JAN SWAMMERDAM
1637–1680

Born in Amsterdam, the Netherlands, Jan Swammerdam qualified as a doctor but dedicated his life to the microscopic study of insects. His main achievements were his discovery of red blood cells and also metamorphosis, which he proved by dissecting an egg, larva, pupa, and adult insect under the microscope and showing that they were simply different forms of the same animal. Swammerdam went on to divide insect development

into four main types, three of which exist in modern classification. But his successes did not impress his father, who withdrew his support. Destitute and depressed, Swammerdam died of malaria aged 43.

NICOLAS STENO
1638–1686

Despite a brief scientific career, Nicolas Steno made major discoveries in anatomy and geology. The Danish scientist likened the teeth of sharks to the "tongue stones" found in Cenozoic rocks. This led him to wonder how fossil shark teeth could have become embedded in rock. Steno went on to establish the underlying principles of how strata (rock layers) form, explaining that the lower the level, the older the rock, and that strata begin as horizontal layers that are disrupted over time. In 1667, he abandoned science for religion and became a bishop 10 years later.

JACOB BERNOULLI
1655–1705

Defying his father's wishes by studying mathematics, Jacob Bernoulli became the first of several famed mathematicians in the Swiss Bernoulli family. He became

professor of mathematics at the University of Basel, Switzerland, in 1687, and taught his younger brother Johann. The brothers worked together on various ways of using calculus, but soon became rivals. Bernoulli is best known for his work in probability and especially for his law of large numbers.

DANIEL BERNOULLI
1700–1782

Jacob Bernoulli's brother Johann had a son, Daniel Bernoulli, who studied philosophy, logic, and medicine and was taught mathematics by his father. In 1738, he published *Hydrodynamica*, which included his discovery that if a stream of moving fluid (liquid or gas) speeds up, its pressure drops. This—Bernoulli's principle—is one of the cornerstones of modern aerodynamics. Bernoulli won 10 awards from the Paris Academy of Sciences in his lifetime, one of which he shared with his father, who—frustrated not to be the sole winner—threw his son out of the house.

EMILIE DU CHATELET
1706–1749

A talented physicist and mathematician, Emilie du Châtelet is also remembered by history for being Voltaire's mistress. Born into a wealthy family in Paris, France, du Châtelet learned science and mathematics from a young age, as well as six languages. Her father, a court official, included her when famous scientists visited their home. She wrote a textbook of current ideas in physics, *Institutions de physique*, which included her own work, and translated the whole of Isaac Newton's *Principia Mathematica* into French. She died in childbirth, with Voltaire and a younger lover by her side.

GEORGES-LOUIS LECLERC
1707–1788

The Comte de Buffon, Georges-Louis Leclerc, caused a stir when he noted the similarities between humans and apes and suggested that living things could be modified by environmental changes such as migration, leading to similar species existing in different places. He also argued that Earth was much older than biblical explanations allowed. Inspired by his work at the Jardin du Roi (the royal botanical gardens in Paris), the French naturalist tried to explain all of nature in his encyclopedia of natural history, but he only published 36 of the proposed 50 volumes before his death.

JOSEPH BLACK
1728–1799

British chemist and physician Joseph Black was born in Bordeaux, France, and studied the arts, then medicine, at the University of Glasgow, Scotland. As part of his 1754 doctoral thesis, Black discovered carbon dioxide, which he called "fixed air." After continuing his studies in Edinburgh, he returned to Glasgow to teach chemistry and began his experiments on the effects of heat on liquids. He found that once a pan of water on a stove reaches boiling point, the temperature of the water stops rising and the heat instead turns the water into vapor. He called this "latent" (hidden) heat. Black rarely published his findings, preferring to announce them to his students during his lectures.

JAN INGENHOUSZ
1730–1799

In 1779, Dutch-born physician Jan Ingenhousz submerged a plant in water and watched as oxygen bubbles formed on the underside of its leaves. He discovered that oxygen formed only on the green parts of the plant that were exposed to sunlight (in a process now known as photosynthesis). He also found that in the dark, plants emit low levels of carbon dioxide. Ingenhousz worked as a doctor in London, England, and was an early advocate of vaccination against smallpox. He even traveled to Vienna to inoculate the Austrian empress Maria Theresa and her family.

JEAN-BAPTISTE LAMARCK
1744–1829

After an injury forced him to leave the French military, Jean-Baptiste Lamarck went to work at the royal botanical gardens in Paris. There, he wrote *Flore Française*, a book about plants. When the gardens became a natural history museum, Lamarck studied invertebrates and came up with the first major theory of evolution. He suggested that over time, living things tend to progress, and that organisms change in response to their environment (for example, by growing longer legs to wade in water). These acquired characteristics are passed on to their offspring. Lamarck's theory was unpopular, and he died in poverty.

PIERRE-SIMON LAPLACE
1749–1827

Born in Normandy, France, Pierre-Simon Laplace developed an interest in mathematics while studying theology in Caen. He moved to Paris, where he secured a teaching position at the Ecole Militaire. In 1773, he began his major work: proving that gravity will keep the solar system stable in the long run and that the self-correcting "perturbations" noted by Isaac Newton (and put down to divine intervention) are, in fact, also made by gravity. As well as proving Newton's theory, Laplace came up with the idea that the solar system formed from a rotating cloud of hot gases.

CHRISTIAN KONRAD SPRENGEL
1750–1816

German botanist Christian Konrad Sprengel achieved little recognition in his lifetime. It was not until 1841 that his work on how fertilization occurs in plants was brought to light by Charles Darwin. Sprengel worked as a teacher in Spandau, Prussia (now Germany), but was dismissed for spending too much time studying plants. He moved to Berlin and, in 1793, published his observations, including the fact that insects carry pollen from the male parts of flowers to the female parts, and that these insects are attracted by the color of a flower's nectaries (nectar-producing organs).

JOSEPH FOURIER
1768–1830

Before settling on a career as a mathematician, Joseph Fourier trained as a priest, dabbled in politics, and traveled with Napoleon to Egypt to study Egyptian relics. The Frenchman made his name with his study of heat flow, in which he showed that a wave of any shape can be represented by adding together simpler waves called sines and cosines. Fourier analysis is now used in many areas, including electronics. He also identified the greenhouse effect, realizing that Earth is warm because gases in its atmosphere trap heat from the Sun and prevent it from escaping.

SCIENCE AND INDUSTRY

1800–1895

MICHAEL
FARADAY

British physicist and chemist Michael Faraday was one of the most influential scientists of the 19th century. His multiple, unprecedented contributions to the fields of electricity and electromagnetism paved the way for an electrical revolution that would change the modern world.

Michael Faraday was born in a poor area of south London, and at age 13 was apprenticed to a bookbinder. As well as learning the art of bookbinding, he read avidly, particularly scientific texts, and developed a thirst for scientific endeavor. After attending a lecture by British chemist Humphry Davy in 1812, Faraday wrote up his copious lecture notes, bound them, and sent them to Davy. Davy offered him a job as his assistant at the Royal Institution (RI) the following year, which set Faraday on a course of scientific inquiry that would revolutionize the world.

At the Royal Institution

Early in his time at the Royal Institution, Faraday accompanied Davy on a tour of Europe, during which he met many high-profile scientists and greatly furthered his scientific education and curiosity. Despite his lack of formal education, Faraday swiftly progressed from assisting Davy to conducting his own experiments. Under Davy's mentorship, he wrote a chemistry manual, discovered new organic compounds (notably

> "One of the greatest scientific discoverers **of all time.**"

Ernest Rutherford, 1931

MILESTONES

ROYAL INSTITUTION
Accepts a job with Davy in 1813 at the Royal Institution. Becomes professor of chemistry in 1833.

KEY DISCOVERIES
Invents the world's first electric motor in 1821 and discovers how to generate electricity 10 years later.

CHEMICAL WORK
Discovers benzene in 1825; in 1834, publishes laws of electrolysis (electrically splitting chemicals).

LIGHT CONNECTION
In 1845, demonstrates for the first time the revelatory link between light and magnetism.

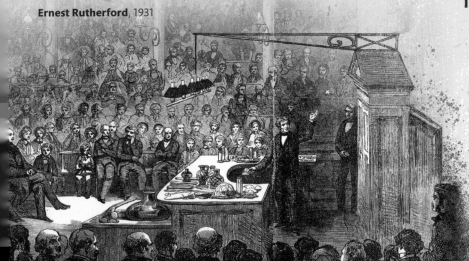

Faraday carried out his research in conjunction with giving regular lectures at the Royal Institution for wide audiences. He hosted 19 public lectures, which were frequently attended by members of the Royal Family.

was run through the wire, the magnet would move around it in a circle. His results, published in 1821, proposed that electromagnetic energy could produce continuous motion: Faraday had created the world's first electric motor.

A decade later, Faraday made an even more significant discovery—that moving a magnetic field can cause electricity to flow through a conductor. In a groundbreaking experiment in which a copper disk was spun between the two poles of a magnet, Faraday generated a steady electric current and produced the first dynamo, or electric generator. This invention had huge implications: during the early 19th century, electricity was the sole authority of scientists and had no real practical application. But Faraday had discovered a means of generating electricity without a battery, and in vast quantities. His method remains the basis on which modern power plants operate.

benzene), and became the first to liquify a permanent gas (one that was believed to be impossible to liquify). Yet it was the field of physics that captured Faraday's imagination and that inspired his key contributions to science.

Discoveries in electromagnetism
In 1820, the Danish scientist Hans Ørsted found a link between electricity and magnetism—namely that when a current is passed through a wire, it produces a magnetic field around the wire. This prompted Faraday's first great discovery. By suspending an electric wire into a cup of mercury with a magnet at the bottom, Faraday demonstrated that if a current

Setting the modern stage
Faraday was a visionary who proved his theories via experimentation, such as the 1845 experiment in which he discovered the effects of magnetism on light. However, some of his ideas proved too radical, including his theory of the "unity of forces," which stated that magnetism, light, electricity, and gravity

TOOK 10 YEARS TO SUCCESSFULLY CONNECT MAGNETISM TO ELECTRICITY IN 1831

SPENT 54 YEARS WORKING AT THE ROYAL INSTITUTION

LAID THE BASIS FOR FUTURE DEVELOPMENTS IN THE FIELD OF ELECTROMAGNETISM

are all manifestations of the same force—an idea that inspired James Clerk Maxwell's electromagnetic field theory (see pp.148–151). His work launched the electrical revolution and paved the way for modern physics. Faraday also set up a series of lectures at the Royal Institution, known as the Christmas Lectures, which continue to inspire new generations today.

> **"Nothing** is too wonderful **to be true,** if it be consistent with the laws of **nature."**

Michael Faraday, 1849

Faraday's electricity generator enabled the development of power plants and the widespread use of electricity, inspiring many modern technologies.

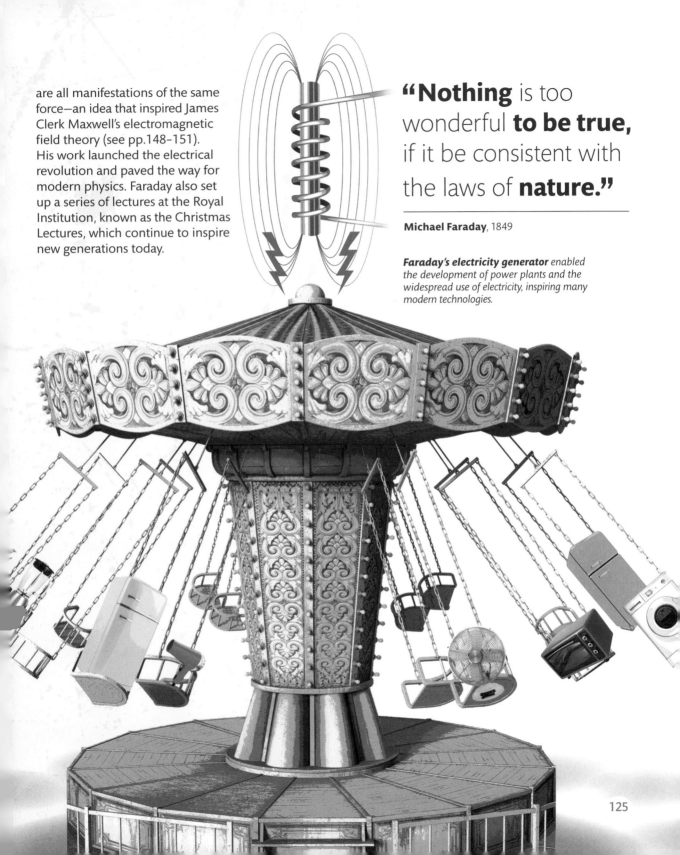

CHARLES
BABBAGE

Charles Babbage was a mathematician and computer pioneer. He designed huge calculating machines, the last of which—his Analytical Engine—is considered to be the forerunner of the modern computer. A lifelong inventor with many interests, he also promoted scientific societies and industrial mechanization.

MILESTONES

DIFFERENCE ENGINES
Begins designing his first calculating machine in 1819. Develops this into the second version in 1847.

GOLD MEDAL
Wins Astronomical Society's Gold Medal in 1824, for inventing a machine to calculate logarithms.

PRESTIGIOUS CHAIR
Made Lucasian Professor of Mathematics at Cambridge University in 1828, but does not give lectures.

MEETS LOVELACE
Becomes friends with Ada Lovelace in 1833. She later sees the potential of his Analytical Engine.

FINAL MACHINE
Designs most of his Analytical Engine by 1838; continues to refine it but never builds it.

Born in London to a wealthy banking family, Charles Babbage displayed a talent and passion for mathematics early on. As a child, he suffered from prolonged ill health, which disrupted his schooling. But he read widely and had learned much about modern mathematics by the time he entered Cambridge University in 1810. He was not impressed with the syllabus there, and this drove him to set up the Analytical Society in 1812, which aimed to promote innovative developments in mathematics.

Babbage was known for his provocative personality. Despite being a top mathematics student at his Cambridge college, he transferred to another one and graduated from there in 1814. He later became influential in scientific circles, and in 1816, he was elected a Fellow of the Royal Society of London. In 1820, he helped found the Royal Astronomical Society.

Manuals and machines for calculating
In the early 1820s, engineers, navigators, and mathematicians used logarithm tables for complex calculations. These tables were devised, copied, and typeset by hand—a process that was vulnerable to errors.

Babbage's Analytical Engine was designed to be programmed using cards punched with holes. It is widely recognized as the precursor of the computer.

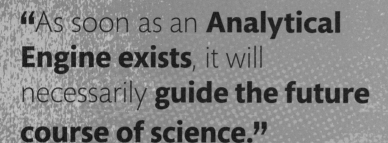

> "As soon as an **Analytical Engine exists**, it will necessarily **guide the future course of science.**"

Charles Babbage, 1864

Invited by the Astronomical Society to lead a project to improve a set of these tables for *The Nautical Almanac*, Babbage found that they contained a large number of mistakes. In 1821, he exclaimed, "I wish to God these calculations had been executed by steam!" He determined to create a foolproof calculating machine. Babbage presented his first design for a Difference Engine in 1822. Powered by cranking a handle, it would formulate and print the mathematical tables by making complex calculations using a process of repeated addition. Babbage secured initial government funding and began work with his engineer, Joseph Clement, to build a prototype: Difference Engine No.1. A small section, about one-seventh of the whole, was finished in 1832, and Babbage liked to demonstrate it to

The Difference Engine No.1 is one of the earliest conceptions of a general-purpose computer. Babbage aimed to mechanize calculation, reducing the risk of human error.

guests. But then, following an argument with Clement about the cost of tools, work on the prototype stopped.

A decade later, Babbage refined his designs to create the Difference Engine No.2. This would need only a third of the parts of the earlier prototype and would be able to calculate numbers up to 31 digits long. However, this time no funding was provided, and the machine was not

constructed in Babbage's lifetime. More than a century later, in 1991, the London Science Museum completed his vision by building a full-scale working model.

The Analytical Engine

In 1833, Babbage started a new project, his Analytical Engine. A more advanced design than the Difference Engine, it would be able to add, subtract, multiply,

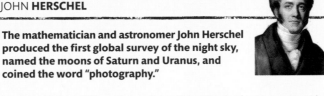

JOHN **HERSCHEL**

The mathematician and astronomer John Herschel produced the first global survey of the night sky, named the moons of Saturn and Uranus, and coined the word "photography."

Herschel (1792–1871) studied mathematics at Cambridge University with Babbage and co-founded the Analytical Society with him. He also helped found the Royal Astronomical Society and served as its president three times. He discovered more than 500 nebulae and 3,000 double stars in the northern hemisphere. In 1821, Herschel was awarded the Royal Society's Copley Medal for his contributions to mathematical analysis. In 1834, he went to South Africa, set up the first observatory in the southern hemisphere, and completed his catalog of the Earth's skies. He was also the first to capture a permanent photographic image, in 1839.

THE **BRITISH GOVERNMENT** CONTRIBUTED **$22,100** TO THE CONSTRUCTION **OF THE DIFFERENCE ENGINE NO.1**

THE **ANALYTICAL ENGINE** COULD STORE **1,000 50-DIGIT NUMBERS:** MORE THAN ANY **PRE-1960S' COMPUTER**

> ## "Babbage has devoted ... years to the **realization of a gigantic idea.**"

Luigi Menabrea, 1842

and divide. It was to be programmed using punch cards held together by string, an idea Babbage copied from the French weaver Joseph Jacquard, who used such punch cards to control the patterns on his looms.

The design for the Analytical Engine had a similar logic structure to that of modern computers. It featured a "store" where intermediate results could be held and a "mill" for arithmetic processing. It could repeat an operation a specified number of times (a process called looping, or iteration) and, depending on the result of an operation, it could choose alternative actions (lateral branching). It could also print results on paper, or as a "stereotype," impressed on plaster of Paris to form a mold from which a printing plate could be made. Although the Analytical Engine was never built, it represented a crucial development from the basic arithmetical processing of the Difference Engine and is widely

regarded as the forerunner of the modern computer. Babbage himself did not publish much about the potential of the Analytical Engine, but when his friend Ada Lovelace (see right) translated a paper about it in 1842, Babbage asked her to expand on the original article, "as she understood the machine so well."

A man of many talents

Babbage was a true polymath. He wrote more than 80 scientific papers on subjects ranging from solar eclipses to geology and statistics; from decimalization to diving bells and lighthouses. He designed a prototype ophthalmoscope (for looking at the retina of the eye) and a "cowcatcher": a metal frame at the front of a train to clear obstacles from railroad tracks.

Babbage's ideas about mechanization also went further than his calculating machine. Inspired by the concept of error-free efficiency, he had suggestions to make about human productivity. In 1832, he wrote *On the Economy of Machinery and Manufactures*, which proposed that industrial manufacturing be improved by mechanization. He also suggested a more streamlined division of labor. This idea is still known as the Babbage principle.

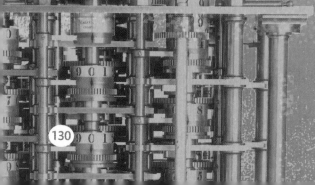

The mathematician and writer Ada Lovelace saw the potential of Babbage's calculating machine. Her instructions for it, written in algebraic code, are regarded as the first-ever computer program.

Born Ada Byron, Lovelace was the only legitimate child of the poet Lord Byron, who was briefly married to her mother, Anne Isabella Milbanke. Aged 17, Ada's tutor, Mary Somerville, introduced her to Charles Babbage. Sharing a passion for mathematics and machines, the two became friends. In 1842, Babbage asked Lovelace to translate a paper about his Analytical Engine, written by the Italian engineer Luigi Menabrea. She added explanatory footnotes to the paper, tripling its length. In these, she set out how to code formulae as instructions for the calculating machine. Lovelace realized that if machines could manipulate numbers, they could also manipulate symbols and process algorithms. Her notes inspired Alan Turing's work on modern computers in the 1940s.

1815–1852

MILESTONES

FLYING ENGINE
In 1828, aged 12, she decides she wants to fly and designs a steam-powered flying machine.

EARLY INFLUENCES
Learns mathematics from her tutor, Mary Somerville, who in 1833 introduces her to Babbage.

A KEEN INTELLECT
Presented at Court in 1833; her brilliance is recognized. She starts to correspond with scientists.

COMPUTER PROGRAM
"Sketch of the Analytical Engine ... with Notes by the Translator" is published in 1843, with her own code.

ANNUAL MEMORIAL
Commemorated on the yearly Ada Lovelace Day from 2009, which celebrates women in science.

ADA LOVELACE

CHARLES DARWIN

One of the most famous naturalists in history, Charles Darwin had no academic training in zoology but was a passionate scientist at heart. On his voyage on HMS *Beagle*, he made careful records of numerous species, which gave him the ideas and the evidence for his theory of evolution by natural selection.

Charles Darwin was born in Shrewsbury, England, the son of a doctor. His mother, the daughter of the famous potter Josiah Wedgwood, died when Darwin was 8, and he was sent to boarding school. He went on to study medicine at Edinburgh University, but left because he found surgery too unpleasant. At his father's wish, he then went to Cambridge University in order to become a clergyman but, although he passed his exams, he spent much of his time studying natural history. In 1831, his friend John Henslow, professor of botany at Cambridge, offered him the chance to join HMS *Beagle* on an expedition that set out to survey the coastline of South America. The voyage lasted 5 years and was to change Darwin's life.

During the trip, Darwin devoted himself to collecting specimens, making detailed drawings, and reading. He was intrigued by the fantastic diversity of species he saw and started to think about how it had developed. He was especially captivated by the unusual wildlife of the Galápagos Islands. Noticing that some species on different islands—

"It is always advisable to **perceive** clearly **our ignorance.**"

Charles Darwin, 1872

MILESTONES

VOYAGE OF A LIFETIME
Embarks from Plymouth on HMS *Beagle* in 1831. Arrives at the Galápagos Islands in 1835.

LOSS OF FAITH
His young daughter Annie dies in 1851, an event that undermines his already-faltering Christian faith.

SEMINAL PAPER
Produces his first paper on evolution, which is presented to the Linnean Society in 1858.

EVOLUTION THEORY
Publishes *On the Origin of Species* in 1859, causing a furor in a conservative society.

FINAL PUBLICATION
Claims that humans are descended from apes in his 1871 landmark book *The Descent of Man*.

Darwin saw marine iguanas and Galápagos finches, species unique to the Galápagos Islands. The highly specialized life forms found there inspired his initial discovery.

mockingbirds, finches, and giant tortoises, for example—shared very similar characteristics, he began to think that in each case they may have shared a common ancestor.

On his return to the UK, Darwin continued to develop his ideas. He was strongly influenced by *An Essay on the Principle of Population* by Thomas Malthus, which predicted that human overpopulation would lead to a struggle for survival due to limited food resources. Darwin wondered if this would apply to animals, too.

Competition and evolution

By 1838, Darwin had formulated his theory of evolution, but he knew it would cause outrage, as it contradicted the Christian view of creation. He was cautious about publishing and instead spent the next 20 years gathering more supporting evidence for his theory. Then, in 1858, he received a manuscript from Alfred Russel Wallace (see box), who had come up with a similar idea. A year later, Darwin published *On the Origin of Species*. The book sold out on the first day of publication.

Darwin's theory of evolution by natural selection was based on his realization that there is always some variation between all the individuals in a species and that more individuals are born than survive. He suggested that, in the competition for survival, it is the characteristics of an individual that make the difference and that only the fittest survive. For example, those with thicker fur are far more likely to

survive in a cold climate and go on to produce offspring. If a useful trait, such as thick fur, can be inherited, more of the next generation will possess that trait. Over time, such small changes add up to a large and noticeable difference.

Although Darwin made sure not to discredit Christian beliefs in his book, it was still a profound shock to Victorian society. His work commanded huge respect nonetheless. When he died, Darwin was given a state funeral and was buried in Westminster Abbey.

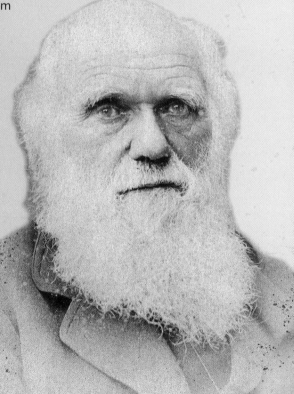

Darwin noticed that the shapes of the beaks of the Galápagos finches varied in accordance with their diets. He proposed that the finches were linked by one common ancestor from which new species evolved as they adapted to the conditions and food of their inhabited islands.

GREEN WARBLER-FINCH

COMMON ANCESTOR

ALFRED RUSSEL **WALLACE**

Darwin and Alfred Russel Wallace devised theories of evolution at the same time.

A British botanist, Wallace (1823–1913) spent years documenting species in the Amazon and the Malay Archipelago. Not blessed with luck, he lost his entire Brazilian specimen collection when his return ship caught fire in 1852. Although eclipsed by Darwin in the history books, his name is preserved most notably in the "Wallace Line" in Indonesia, which marks a division between species of Asian origin and those of Australian origin.

SMALL TREE-FINCH

MEDIUM GROUND-FINCH

"This preservation of **favorable variations** and the rejection of **injurious variations**, I call **natural selection."**

Charles Darwin, 1859

LARGE GROUND-FINCH

"FOR MY OWN PART I WOULD AS SOON BE DESCENDED FROM THAT HEROIC LITTLE MONKEY, WHO BRAVED HIS DREADED ENEMY IN ORDER TO SAVE THE LIFE OF HIS KEEPER ... AS FROM A SAVAGE WHO DELIGHTS TO TORTURE HIS ENEMIES."

Charles Darwin
The Descent of Man, 1871

◄ *HMS* **Beagle** *anchors at Tierra del Fuego, Argentina, c.1832.*

JAMES PRESCOTT
JOULE

A brewer by trade and a keen amateur scientist, British physicist James Prescott Joule investigated the nature of heat and the transfer of energy. He was renowned for experiments that proved heat was a form of energy and for confirming one of the most fundamental ideas in physics: the law of conservation of energy.

The owner of a successful brewery in northern England, James Prescott Joule fostered his interest in science by conducting experiments into the nature of heat in its laboratories. Although lacking formal training, he was a talented experimenter who made key discoveries during his career.

Experimental science

The results of Joule's early investigations into the relationship between electrical resistance and heat became known as Joule's first law. He then examined the direct relationship between mechanical work and heat, devising an experiment to mechanically turn a paddlewheel in water to raise its temperature. Joule concluded that the mechanical energy was converted into heat and argued that energy is never lost, but merely changes form. This groundbreaking theory became the foundation of the law of conservation of energy. He also discovered the Joule-Thomson effect with Lord Kelvin in 1852, which examined expansion and temperature changes in gases and led to pioneering work in refrigeration.

"Energy may be **converted** into **heat**; **heat** into energy."

James Prescott Joule, 1847

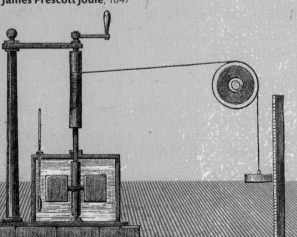

Joule's most famous experiment used falling weights (right) to mechanically churn and heat water (left). He noted that the falling weights generated energy that converted into heat as the paddles created friction in the water.

GREGOR MENDEL

The Augustinian monk Gregor Mendel is widely regarded as the founder of modern genetics. He discovered the laws of heredity and realized that family traits are inherited through units that are now known as genes.

Johann Mendel was born to peasant farmers in a village in what is now the Czech Republic. He initially trained in mathematics and philosophy before entering the priesthood as a way of furthering his education. Renamed Gregor when he joined St. Thomas's Abbey in Brno, Mendel was sent to Vienna University to complete his studies, after which he returned to teach at the abbey.

Mendel was interested in plant breeding and heredity. Inspired to help the monks improve their crops, he focused on peas. His mathematical background helped him to design his experiments: he looked at "either/or" traits, such as seed color, flower color, and plant height, which made it easier for him to interpret his results.

Generations of pea plants
In one example, Mendel crossed green-seeded and yellow-seeded peas, and all the seeds in the next generation were yellow. He then crossed two of their offspring with each other and found that the

> "We have now the means of beginning **an analysis of living organisms.**"
>
> **William Bateson**, 1913

MILESTONES

FURTHER EDUCATION	FIRST EXPERIMENTS	SEMINAL PAPER	HIGHER CALLING
Works as a teacher while a friar, then goes to study at the University of Vienna in 1851.	Authorized to study the St. Thomas's Abbey garden in 1854; begins to cultivate peas in 1856.	Publishes his *Experiments in Plant Hybridization* in 1866, but its significance is unappreciated.	Becomes Abbot of St. Thomas's in 1868, and puts aside his scientific research.

1822–1884

second generation contained yellow-seeded and green-seeded plants in the exact ratio of 3:1. Mendel saw that "either/or" traits were passed on in exact proportions. He called this the "law of segregation." He also realized that these characteristics were determined by discrete "particles," now called genes, that occurred in pairs and were formed at fertilization because one gene was inherited from each parent plant.

Dominant and recessive genes

As Mendel discovered, there are different versions of a gene for any particular trait. Some are stronger ("dominant") and some weaker ("recessive"). Dominant versions override recessive ones, so when different versions are inherited, only the dominant trait appears in the next generation. With Mendel's peas, the gene for yellow seeds was dominant. As the offspring of his first cross inherited one "yellow" gene and one "green" gene, all produced yellow seeds. But when two of these plants were crossed, one-quarter of their offspring would inherit a "green" (recessive) gene from each parent and have green seeds, while the rest were yellow. This hereditary mechanism explains why some traits can skip generations: for example, a red-haired parent might have a dark-haired child but a red-haired grandchild.

It was fortunate that Mendel studied peas, as their genetic mechanism is very simple. The process in many other organisms is more complex, although still based on the same rules. Most characteristics are governed by a combination of genes, resulting in a more variable outcome (height in humans, for example). Moreover, some genes for different traits are linked together and are therefore inherited together.

Mendel published his findings, but they were ignored for more than 30 years. Around 1900, they were rediscovered, and he received the credit he deserved.

In his pea-breeding experiments, Mendel found that some genes could be carried without being expressed as physical traits. In this pea "family tree," the first generation all carry the gene for red seeds, and the red-seeded trait becomes visible when the genes recombine in the second generation.

HUGO **DE VRIES**

The Dutch botanist de Vries independently reached the same conclusions as Mendel about heredity, and brought genetics to the attention of the scientific world.

The young de Vries (1848–1935) was passionate about plants and was also an admirer of Darwin's theory of evolution. Intrigued to learn more about how evolution worked, he carried out plant-breeding experiments on evening primrose and used the word "mutation" for the spontaneous variations that sometimes occurred. Further experiments led him to the same results as Mendel, which he published in 1900. De Vries also coined the term "pangene" for the unit of heredity, which later became "gene."

"Recessive traits ... **reappear unchanged in their progeny.**"

Gregor Mendel, 1866

Trained in chemistry and biology, Louis Pasteur was instrumental in establishing the field of microbiology. He discovered that microorganisms cause infectious disease and developed the first vaccines for rabies and anthrax. He also invented a method of sterilizing food, known as "pasteurization."

Louis Pasteur was born and raised in a small town in the Jura region of France. His father, like generations before him, was a leather tanner. A talented artist, Louis spent much of his time drawing portraits in pastels. On his second attempt, he passed the entrance exams for an elite college in Paris, and in 1847 submitted PhD theses in both physics and chemistry. The next year, he became professor of chemistry at the University of Strasbourg and made his first scientific discovery: that some molecules exist in two mirror-image forms.

Fermentation, pasteurization, and microbes

In 1854, Pasteur became Dean of Sciences at the University of Lille. He became interested in a problem that was afflicting winemakers: the fermentation process often went wrong, spoiling the wine. Pasteur discovered that fermentation is caused by living yeast microbes and that souring occurs when the wine is contaminated by bacteria. In 1865, he patented a method for preventing such "diseases" of wine. The liquid was heated briefly to 140–212°F (60–100°C), which killed off the bacteria without affecting the taste. This process, which also worked on beer and milk, became known as pasteurization.

Pasteur's research went further than solving the wine problem. He also answered the question of how microbial contamination occurs. In the 19th century, most scientists believed in the concept of

News spread quickly of the first successful rabies vaccination, administered to the 9-year-old Joseph Meister, and people soon flocked to Pasteur for treatment.

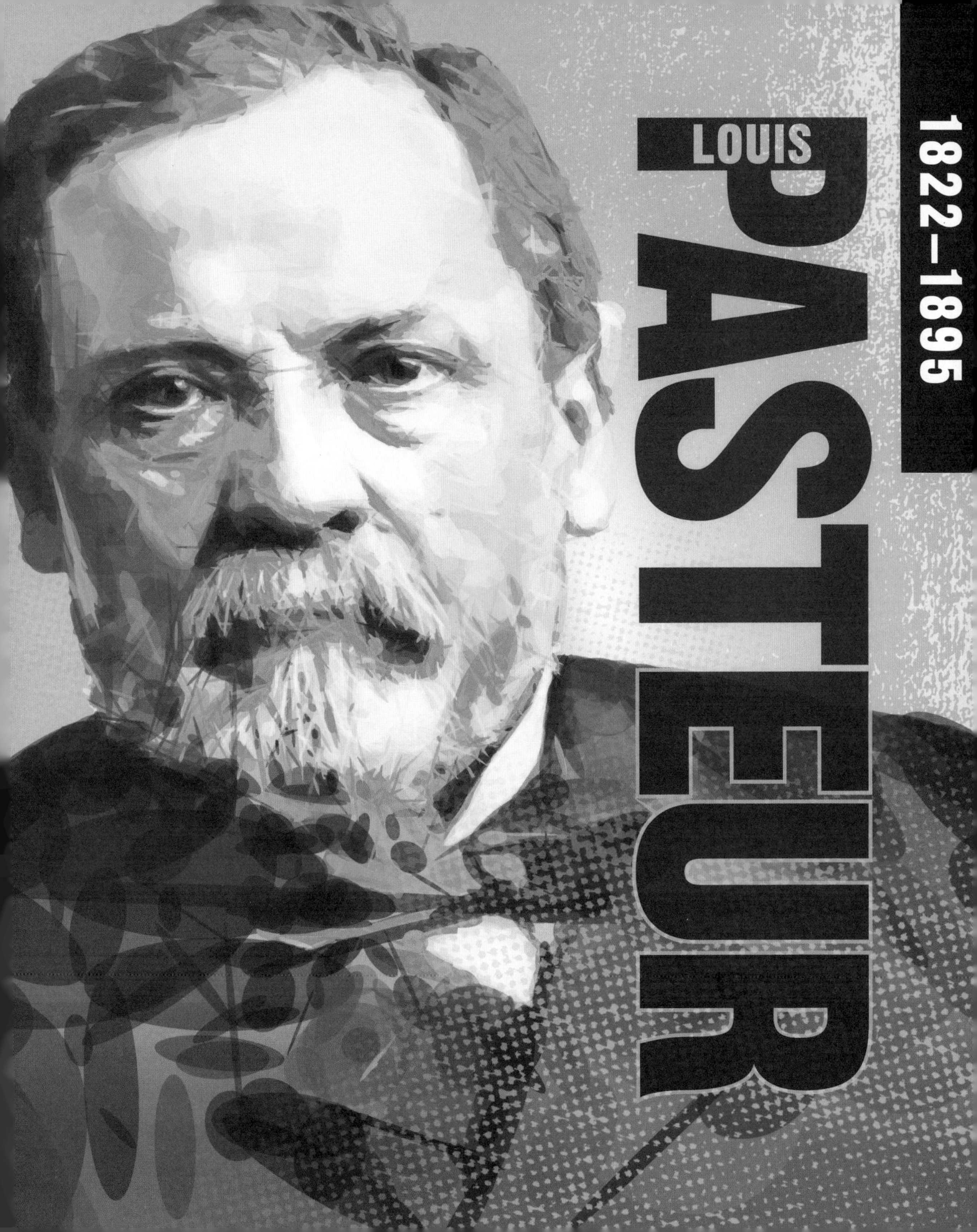

LOUIS PASTEUR

1822–1895

"abiogenesis": that life could generate spontaneously from nonliving material. For example, maggots were thought to just appear by themselves in rotting meat. Likewise, infectious diseases were thought to be caused by "miasma" (bad air) rising from decaying matter.

In a famous experiment, Pasteur demonstrated that the air is filled with microbes, and that these will colonize any exposed surface or liquid. First, he filtered air through cotton, examined the cotton through a microscope, and found that it contained the type of microorganisms associated with decaying food. Next, he heated a glass flask of nutrient-rich broth to sterilize it. He then softened the neck of the flask with heat and bent it down and up again into a "swan neck" S-shape. This meant that nothing in the air above the flask could fall directly down onto the liquid in the flask. When the broth cooled, no microbes reappeared in it. However, when Pasteur tilted the

"Everything is clear if its cause be known."

Louis Pasteur, 1878

neck of the flask so that the outside air could reach the broth, it became contaminated with microbes once more.

Germ theory
After his help with the wine industry, the French government asked Pasteur if he could cure a disease that was destroying silkworms. Pasteur succeeded, having found that it was caused by two different types of parasitic microbe. Now convinced that illness was brought about by certain microbes (germs), he devoted himself to learning more. First, he studied chicken cholera. He cultured

ROBERT **KOCH**

The German physician Robert Koch was the founder of bacteriology. He discovered the life cycle of anthrax and identified the bacteria that cause cholera and tuberculosis.

After graduating from medical school with flying colors, Koch (1843–1910) soon developed a passion for laboratory work on microorganisms. He devised techniques for culturing bacteria and discovered that the anthrax bacterium produces a type of spore, which can remain dormant in soil. He is particularly recognized for his work on isolating specific microbes: his four "Koch's postulates" established the key principles for identifying the microbe that causes a disease. In 1905, Koch won the Nobel Prize in Physiology or Medicine for his research into tuberculosis. Once a colleague of Louis Pasteur's, the two men later became archrivals.

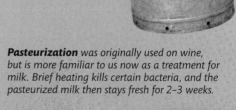

Pasteurization was originally used on wine, but is more familiar to us now as a treatment for milk. Brief heating kills certain bacteria, and the pasteurized milk then stays fresh for 2–3 weeks.

cholera microbes in his laboratory. When inoculated (infected) with these microbes, his chickens became ill and died. But on one occasion, in 1879, the microbe culture was abandoned for a month while Pasteur went on vacation.

When Pasteur inoculated the birds with it, they became ill but recovered. He then inoculated them with fresh cultures, which had killed other birds, and they survived. Pasteur realized that his chickens had developed an immunity to cholera as a result of exposure to the weakened culture. What he had created was a chicken cholera vaccine.

Pasteur extended his studies to anthrax in cattle and sheep, then began to work on a vaccine for rabies. In 1885, he injected his vaccine into a 9-year-old boy called Joseph Meister, who had been bitten by a rabid dog. Joseph survived, and Pasteur became a national hero.

The realization that microorganisms cause disease and can spread through the air or by direct contact revolutionized medicine. It led not only to vaccination regimes but also to better basic hygiene practices, crucial to preventing infection.

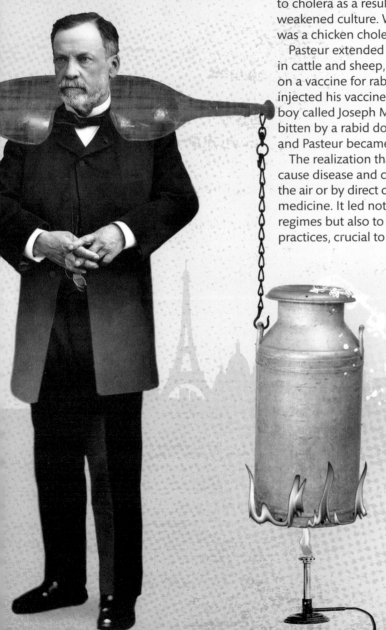

GAVE **MEISTER** **13** VACCINE INJECTIONS OVER **10** **DAYS;** **MEISTER** **RECOVERED**

DEVISED METHODS TO **PROTECT** **HUMANITY** FROM **2** **DEADLY** DISEASES— **ANTHRAX** AND **RABIES**

Nineteenth-century British physicist James Clerk Maxwell's theory of electromagnetism proved that electricity, magnetism, and light are all different aspects of the same underlying phenomenon: an idea that remains of fundamental importance.

James Clerk Maxwell was an only child, the son of a wealthy Edinburgh lawyer. His mother, who tutored him at home in his early years, died when he was 8 years old. The young Maxwell was extremely clever and inquisitive and intrigued by how things work. At the age of 14, he published his first scientific paper, on the subject of geometry. He also memorized the Bible and remained a devout Protestant all his life.

Maxwell went to Edinburgh University at the age of 16 to study physics (then called "natural philosophy") and mathematics. Finding the taught courses unchallenging, he began to conduct his own experiments in his spare time and published two more papers. Aged 19, he went to study mathematics at the University of Cambridge, and 5 years later became a Fellow of Trinity College. His intellectual interests were many and varied: in 1855, he published "Experiments on color as perceived by the eye." He later created the first color photograph.

The rings of Saturn

At 25, Maxwell took up a professorship at the University of Aberdeen. He also immersed himself in a challenge that had been set for the Adams Prize, a prestigious mathematics prize at Cambridge: to explain Saturn's rings. Scientists could not understand how the rings could

The first-ever color photograph, of this tartan ribbon, was produced by Maxwell in 1861. He made it using red, green, and blue color filters.

"**Every atom** of creation is unfathomable in its **perfection.**"

James Clerk Maxwell, 1873

JAMES CLERK

MAXWELL

1831–1879

remain stable without breaking up. Maxwell spent 2 years working on the problem before showing that the rings could not be solid or fluid, but must be made of innumerable small particles. He won the prize in 1859. His analysis was ultimately proven in 1980, when NASA's Voyager 2 reached Saturn.

In 1860, Maxwell was appointed to the Chair of Natural Philosophy at King's College, London, and his interests broadened further. He greatly advanced the kinetic theory of gases, showing the mathematical relationship between the temperature of a gas and the different speeds of its molecules. However, it is for his work on electromagnetism that he is most renowned.

A grand unified theory

Maxwell was intrigued by the ideas of Michael Faraday (see pp.122–125), who had demonstrated clear links between electricity and magnetism. Maxwell wanted to make mathematical sense

Maxwell's discovery that visible light is an electromagnetic wave (made of separate electrical and magnetic waves vibrating at right angles) was one of the most important advances in physics since Newton's laws.

HEINRICH **HERTZ**

German physicist Heinrich Hertz discovered radio waves and was the first to prove electromagnetic theory by experiment. The unit of frequency is named after him: the hertz.

Hertz (1857–1894) was a gifted young man who worked in a physics laboratory in Berlin. He was invited to enter a competition to verify James Clerk Maxwell's theory, but did not think he was capable. Years later, in 1886, he noticed that electric sparks caused a regular vibration, resulting from fluctuating electric charges. He realized that, if Maxwell was right, these would radiate electromagnetic waves. He designed the necessary apparatus and produced radio waves. In fact, he had made the first radio transmitter.

"Maxwell is the physicist's physicist. He is **the unsung hero of British science.**"

Stephen Hawking, 2010

of Faraday's experimental results. In 1855 and 1856, he proved that electricity and magnetism are actually different aspects of one force: electromagnetism. He then went further—he calculated what the speed of an electromagnetic wave would be if it was produced by an electric current in a conducting loop. The answer was the speed of light. Maxwell realized that light itself must be an electromagnetic wave. Indeed, his equations showed that there could be an infinite number of frequencies of electromagnetic waves, all traveling at the speed of light, and that the light familiar to us ("visible light") is only a small part of this spectrum. Since his time, other kinds of electromagnetic waves have been discovered, including X-rays, microwaves, and radio waves. Maxwell published his theory in 1864, with a series of equations. After his death, these were summarized into just four, now known as "Maxwell's laws." While the mathematics is not easy to comprehend, its significance cannot be understated.

The theory was a huge conceptual leap in the understanding of the fundamental forces of nature—and it paved the way for Einstein's special theory of relativity. Indeed, Einstein (see pp.198–203) is quoted as saying: "One scientific epoch ended and another began with James Clerk Maxwell."

PUBLISHED **FIRST PAPER** AT THE AGE OF **14**

DISCOVERED LIGHT WAVES ARE **200** MILLION TIMES **SHORTER** THAN RADIO WAVES

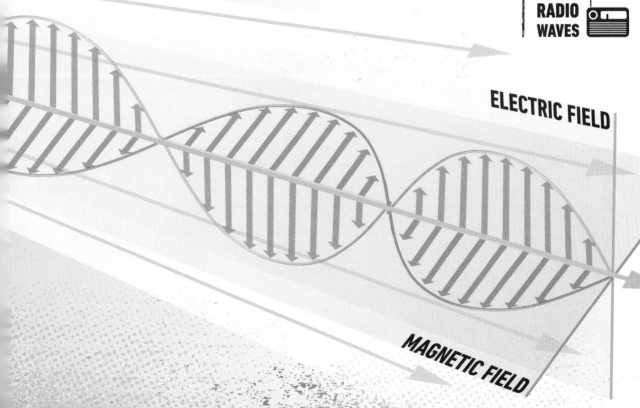

ELECTRIC FIELD

MAGNETIC FIELD

DMITRI
MENDELEEV

Known as the creator of the periodic table, Dmitri Mendeleev was an eccentric chemistry professor who solved the problem of how the elements are related to each other. His classification system advanced science, leading to the discovery of new elements and an understanding of the structure of the atom.

MILESTONES

CHEMISTRY PIONEER

Discusses standardizing chemistry at the first-ever international chemistry conference in 1860.

PERIODIC TABLE

His brainwave comes in 1869, when he devises his periodic table of elements in a grid pattern.

FIRST REFINERY

Conducts research into petroleum and helps to found the first Russian oil refinery in 1879.

WINS ACCLAIM

In 1905, is awarded the Copley Medal and elected to the Royal Swedish Academy of Sciences.

PRIZE NOMINEE

Nominated for the Nobel Prize in Chemistry in 1906, but is rejected, possibly due to a personal grudge.

Dmitri Ivanovich Mendeleev was born into a poor Russian family in a village in Siberia, the youngest of 14 surviving children. When Mendeleev was 13, his father went blind, and his mother had to support the family by running a glass business. Two years later, the factory burned down. Determined that Mendeleev should have a higher education, his mother walked and hitchhiked with him across Russia to St. Petersburg. Tragically, she died just days after he enrolled at the university there.

Mendeleev shone at his studies and later became professor of chemistry. He was an idiosyncratic teacher, known for his disheveled appearance and fiery temper. His career flourished, however, and his textbook *The Principles of Chemistry* was published between 1868 and 1870; it has since been translated into several languages.

One of the key challenges within chemistry at the time was how to classify the elements. Several scientists had identified certain patterns in the way elements behaved—for example, some reacted violently with others, while others hardly seemed to react at all. It was known that

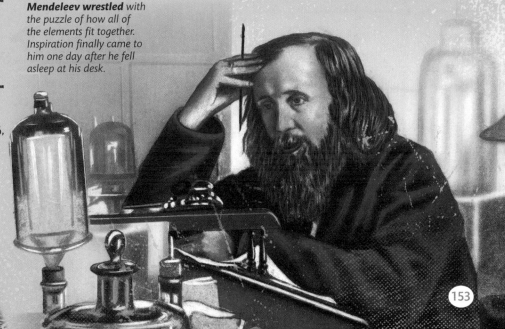

Mendeleev wrestled with the puzzle of how all of the elements fit together. Inspiration finally came to him one day after he fell asleep at his desk.

the atoms of each element have a different atomic mass and, in 1865, the British chemist John Newlands realized that if you listed the elements in order of their atomic mass, there was a recurring pattern, or "periodicity," in their physical and chemical properties. This periodicity occurred every eighth element, so he called it the "law of octaves." However, the pattern seemed to break down after the 20th element, calcium, and there were gaps.

Dreaming up ideas

Mendeleev felt compelled to solve the puzzle. In 1869, he had the idea of writing the name and number of each of the 63 then-known elements on the back of a playing card so that he could experiment with different arrangements. He tried to represent the elements' periodicity with a grid of rows and columns, but it did not quite work.

Working on the problem one night, he fell asleep in front of his cards. He had a dream about playing Solitaire—and woke

up with a new idea: why not try shaping the grid pattern to match the elements' properties, leaving gaps as necessary that would correspond to elements that had not yet been discovered? It worked beautifully. His periodic table comprised seven columns, corresponding to seven groups—elements with "family" traits. An eighth group, the "noble gases," was discovered during the 1890s, and fit neatly onto the table.

One of the most exciting things about Mendeleev's table was that it made it possible to predict the properties of the elements that would fill the gaps simply because of their positions. By 1888, he was proved right: his three gaps had been filled by the new elements gallium, scandium, and germanium.

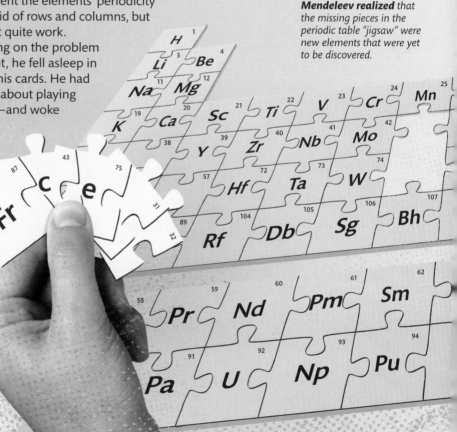

Mendeleev realized that the missing pieces in the periodic table "jigsaw" were new elements that were yet to be discovered.

"Work, **look for peace and calm in work**; you will find it nowhere else."

Dmitri Mendeleev

The periodic table was a game-changer for the science of chemistry. For the first time, it made a connection between the atomic mass of each element and the way it chemically reacted. Mendeleev's crucial role in this was honored nearly a century later, in 1955, when a new synthetic element was named after him: mendelevium.

HENRY MOSELEY

In 1913, 26-year-old Henry Moseley worked out the fundamental property of each element.

Moseley (1887–1915) was a brilliant British physicist and experimentalist. He discovered, using X-ray spectroscopy, that elements were defined by the number of protons in their nucleus (atomic number) rather than their atomic mass. Once the elements were listed in this order on Mendeleev's periodic table, they fit together perfectly, and all the imperfections disappeared.

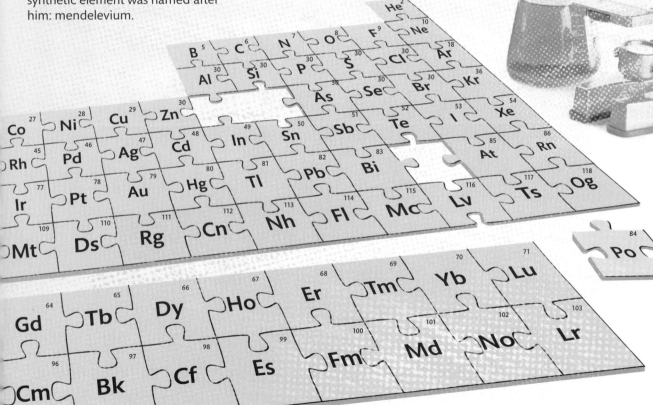

"WILLINGLY OR NOT, IN SCIENCE WE ALL MUST SUBMIT NOT TO WHAT SEEMS TO US ATTRACTIVE FROM ONE POINT OF VIEW OR FROM ANOTHER, BUT TO WHAT REPRESENTS AN AGREEMENT BETWEEN THEORY AND EXPERIMENT ..."

Dmitri Mendeleev

"The Periodic Law of the Chemical Elements", *Faraday Lecture*, 1889

A revered scholar, Mendeleev was honored with numerous titles and medals. ▶

ALEXANDER
GRAHAM BELL

Best known for patenting the first telephone, British-born engineer Alexander Graham Bell also invented the photophone (a wireless communication device), refined the phonograph (which became the record player), and conducted groundbreaking work in the fields of surgery and nautical engineering.

MILESTONES

AWARDED PATENT
Receives a patent in 1876 for his technology that enables speech to be transmitted over wires.

PUBLIC APPEARANCE
Demonstrates his telephone publicly in 1876, making a call to a location 10 miles (16 km) away.

STARTS COMPANY
Founds the Bell Telephone Company in 1877, with financial backing from his father-in-law.

WIRELESS PHONE
Invents the photophone in 1880. The device transmits speech wirelessly using light.

In 1874, Alexander Graham Bell, a teacher to the deaf, began work on a device that, like the telegraph, used wires to transmit speech over long distances. Several other inventors were working toward the same goal—including Elisha Gray, who developed a liquid transmitter that Bell then adapted for his own experiments—but Bell beat them to producing and patenting a practical telephone. After numerous experiments with Thomas Watson, whom he recruited to help engineer his ideas, Bell discovered that the combination of a single steel reed relay vibrating over rotating magnets generated a current powerful enough to reproduce sound over a distance. It was this innovation that made telegraphic speech possible.

After further experimentation, Bell devised an electromagnetic transmitter, which he incorporated into his Butterstamp phone (named for its resemblance to the tool used for making butter pats). Patented in 1876, this was the first commercially viable telephone. But it was not the last of Bell's inventions. In 1877, he improved Thomas Edison's phonograph, which both recorded and replayed sound. He then developed a metal detector that could locate bullets in the human body—a device that was widely used in the Boer War and World War I. He also created the first hydrofoil; co-founded the National Geographic Society; and designed the tetrahedral space frame, forerunner of a construction technique that would revolutionize architecture and engineering in the 20th century.

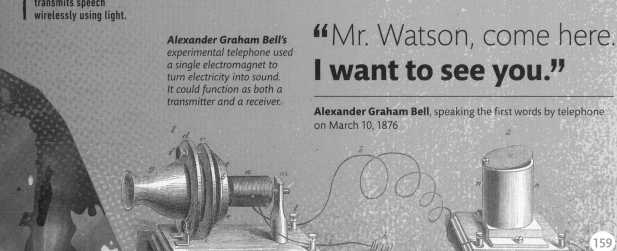

Alexander Graham Bell's experimental telephone used a single electromagnet to turn electricity into sound. It could function as both a transmitter and a receiver.

"Mr. Watson, come here. I want to see you."

Alexander Graham Bell, speaking the first words by telephone on March 10, 1876

DIRECTORY

The flourishing relationship between science and industry during the 19th century fueled the rise of science as a profession. Backed by strong academic institutions, scientists shared information and were able to begin to unify individual discoveries into coherent wider theories.

ALEXANDER VON HUMBOLDT
1769–1859

Born in Berlin, Prussia (modern-day Germany), Alexander von Humboldt was an originator of modern geography and ecological thinking. He was one of the first scientists to study how physical conditions—such as climate, altitude, latitude, and soils—affect the distribution of life. A great explorer as well as a scientist, he spent 5 years in South America, where he helped map how flora and fauna differed between sea level and high altitude in the Andes. He also noted that the rotation of Earth did not affect the direction of ocean currents, and he proposed that Africa and South America were once linked. After returning to Europe, he spent much of his life writing up his findings in a five-volume work called *Kosmos*.

HANS CHRISTIAN ØRSTED
1777–1851

Danish scientist Hans Christian Ørsted made his pivotal discovery linking magnetism and electricity while lecturing at the University of Copenhagen. He was showing how the electric current from a voltaic pile (an early battery) can heat up a wire and make it glow when he noticed that a compass needle standing near the wire moved every time the current was switched on. Further study convinced him that the current produced a circular magnetic field as it flowed through the wire. Ørsted's discovery changed how scientists understood electricity and magnetism, which were previously considered as two separate forces of nature. Further developed by Michael Faraday and James Clerk Maxwell, it led to the modern theory of electromagnetism.

HUMPHRY DAVY
1778–1829

After a brief career studying gases in Bristol's Pneumatic Institution, English chemist Humphry Davy was invited to join the Royal Institution in London. It was here he discovered new elements, including potassium and sodium, through a process that later became known as electrolysis. Davy realized that the electricity produced by a voltaic pile (an early battery) is the result of the chemical reactions happening inside it and that reversing the process would allow electricity to split chemical compounds into the elements from which they were made. Many previously unknown elements were discovered in this way.

FRIEDRICH WOHLER
1800–1882

Born near Frankfurt in what is now Germany, Friedrich Wöhler graduated in medicine but became a distinguished chemist after studying under Jöns Jacob Berzelius in Sweden. His reputation grew first through his purification of aluminum. Then, in 1828, he accidentally created urea from ammonia and cyanic acid. Previously, it was believed that biological, organic compounds such as urea could only be obtained from a living organism.

JUSTUS VON LIEBIG
1803–1873

German chemist Justus von Liebig made many advances in the chemistry of organic substances, which at the time meant substances made in living organisms. Liebig maintained that the secret of organic chemistry lay in knowing that the properties of substances containing just carbon and hydrogen (and sometimes oxygen and nitrogen as well) were as variable as all other elements put together. He later asserted that chemists would one day be able to synthesize sugar, as well as the painkillers salicin and morphine.

RICHARD OWEN
1804–1892

The first person to coin the word "dinosaur," Richard Owen used the term *Dinosauria* (meaning "terrible lizard") to group three fossil finds made in the early part of the 19th century. Unlike ordinary reptiles, he said, dinosaurs stood on erect limbs, and their backbones above the hips were fused together. Owen strongly opposed the views of Charles Darwin and used the dinosaurs as an argument against evolution, claiming they were more advanced than a living reptile.

MATTHEW FONTAINE MAURY
1806–1873

A pioneer in oceanography, Matthew Maury was born in Virginia and spent time in the navy before an accident ended his seagoing days. From 1842–1861, he was the first superintendent of the US Naval Observatory. He devoted himself to charting the winds and currents of the North Atlantic by analyzing old ships' logs. In 1855, he published *The Physical Geography of the Sea*, now credited as the first oceanography textbook.

THEODOR SCHWANN
1810–1882

German physiologist Theodor Schwann laid the foundation of cell theory—the identification of the cell as the basic unit of animal structure. He defined the three structural parts of a cell—the wall, nucleus, and fluid content—and said these parts were present in animal as well as plant cells. In 1839, Schwann published the paper *Microscopic Investigation on the Accordance in the Structure and Growth of Animals and Plants*, in which he famously observed that "all living things are composed of cells and cell products." The Schwann cell—a cell that forms the myelin surrounding the axons of peripheral nerves—was later named after him.

CLAUDE BERNARD
1813–1878

French-born Claude Bernard carried out pioneering work in experimental medicine, including how the body is regulated. His findings provided the foundations for the modern concept of homeostasis—the means by which the body maintains a stable environment that is independent of external changes. He also studied the various processes of digestion—in particular, the roles of the pancreas and liver—and described how chemicals are broken down by digestion into simpler substances and then built up again to make new molecules for body tissues.

WILLIAM THOMSON
1824–1907

Physicist William Thomson, who later became Lord Kelvin, was born in Belfast and raised in Glasgow. His work brought together many areas of physics, and he stated that any physical change is fundamentally a change in energy. He elaborated on the second law of thermodynamics—which states that in any closed system, everything will eventually reach the same temperature—and proposed a temperature scale that was named after him. The Kelvin scale starts at absolute zero (which is -459°F/-273°C) but uses the same units as the Celsius scale. He also oversaw the laying of the trans-Atlantic telecommunications cable.

FRIEDRICH AUGUST KEKULE
1829–1896

Born in Darmstadt, Germany, Freidrich August Kekulé was the main originator of the notion of chemical structure. He was educated in Germany, France, Switzerland, and London before becoming a professor at the University of Bonn. He studied the valency of elements (the maximum number of other atoms they can combine with), and in 1857, he announced that carbon has a valency of 4. His most notable discovery was that the structure of benzene is a symmetrical ring of carbon atoms. Kekulé allegedly "saw" this arrangement after having a dream of a snake biting its own tail.

CARLOS JUAN FINLAY
1833–1915

Cuban physician Carlos Juan Finlay was the first to suggest, in 1881, that yellow fever was transmitted by a species of mosquito. He had studied in Paris from 1860–1861, before returning to Cuba to work as a general practitioner. He became interested in infectious disease and noticed that the *Aedes aegypti* mosquito was often present in houses during yellow fever epidemics. It was 20 years before his ideas were taken seriously. In 1900, researchers on the US Army Yellow Fever Board proved the connection by experimenting on themselves: some bitten by the mosquito went on to develop yellow fever.

AUGUST WEISMANN
1834–1914

German biologist August Weismann is regarded as the most significant evolutionary biologist after Darwin. His lifelong interest in biology began as a boy, and he went on to study medicine. A supporter of Darwin's theory, Weismann proved that inheritance occurs only via sperm and eggs and not from body cells. He did this by cutting off the tails of mice for five generations. All offspring were born with tails, showing that acquired characteristics could not be inherited.

PARADIGM
SHIFTS

1895–1925

Nobel Prize-winning neuroscientist Santiago Ramón y Cajal used microscopes to explore the organization of the nervous system. He was the first to realize that it is composed of independent cells that communicate with one another.

Born in northern Spain, Santiago Ramón y Cajal was a rebellious child with a passion for drawing. He wanted to be an artist, but his father, a practicing doctor who also taught anatomy, persuaded him to go to medical school. Eventually, Cajal's artistic talents and scientific interests merged perfectly in the fast-developing field of neuroscience.

Microscopes and staining
While studying to become a doctor, Cajal used a microscope for the first time. Impressed, he used his savings to buy one and began to view and draw the structure of the muscle tissue that he saw on an increasingly small scale. In 1885, he was given a more modern microscope, and his work became entirely focused on histology—anatomy at microscopic level. Before long, he had started to investigate the nervous system.

Prior to the 1870s, using a microscope to examine nervous tissue had had limitations—low magnification and poor resolution made it hard to determine where one structure ended and another began. But this changed with advances in optical lenses and better versions of the stains that were applied to tissue samples in order to make individual cells more visible.

In 1887, while professor of histology and pathological anatomy at the University of Barcelona, Cajal saw brain tissues that had been stained using

Having modified Golgi's tissue-staining technique, Cajal produced hundreds of illustrations of the human brain and nervous system.

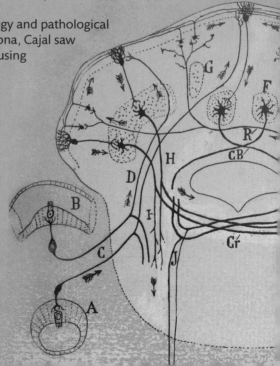

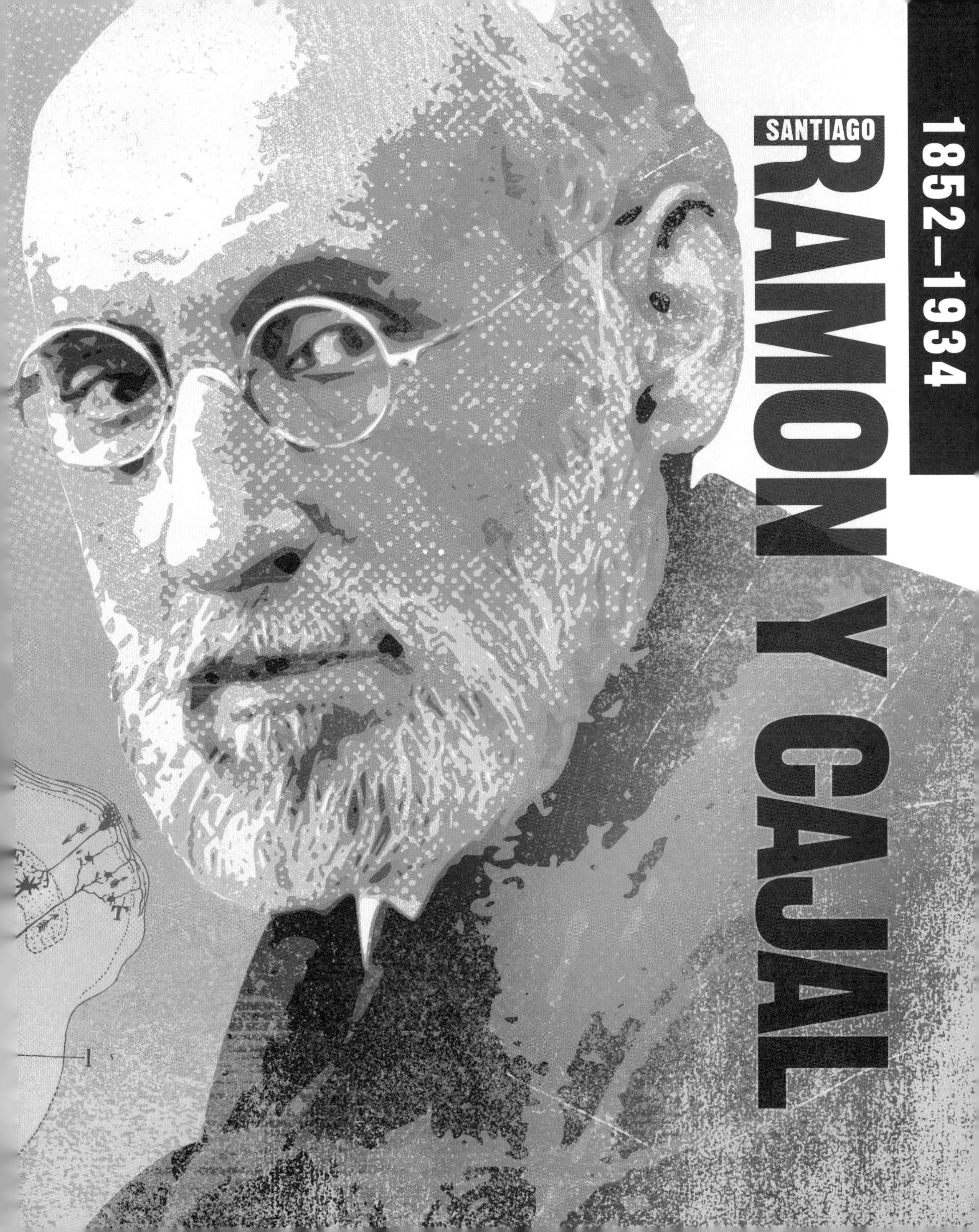

SANTIAGO **RAMON Y CAJAL**

1852–1934

Camillo Golgi's method (see box) and was amazed. It was a career-changing moment for him. "The nerve endings could be seen," he wrote later. "A look was enough. Dumbfounded, I could not take my eye from the microscope."

After trying Golgi's method, Cajal worked to improve it until it allowed for the complete visibility of the nerve cells in the brain, eye, and spinal cord tissue of birds and embryonic mammals. He then drew by hand, in intricate and exquisite detail, the stained cells he saw through his microscope, creating some of the world's greatest scientific illustrations.

Neuron doctrine

Over a 6-year period, Cajal continued to apply his modified staining technique to other kinds of nervous tissue from different animals, including humans. By 1889, he had observed and recorded that the brain and nervous system are composed of billions of individual, independent nerve cells that communicate with each other—through electrical and chemical

transmission—across tiny gaps (now called synapses). Cajal's observations refuted the prevailing theory on the composition of the nervous system—reticular theory—which claimed that the nerve fibers were fused together to form a single, continuous network.

In 1891, German anatomist Wilhelm Waldeyer coined the term "neuron" to mean nerve cell and used Cajal's findings to support a "neuron doctrine"; this states that the basic units of the nervous system are discrete cells (neurons) that have a very specific connectivity that determines how they

signal to each other. By the end of the 19th century, the neuron doctrine had replaced reticular theory. It gained widespread acceptance in the 1950s, when electron microscope images revealed the existence of synapses. Today, it is the model for the structure and function of the nervous system.

In 1906, Cajal and Golgi were jointly awarded the Nobel Prize in Medicine or Physiology for their separate studies on the nervous system; Golgi, however, was a staunch reticulist and still insisted that nerve cells were physically connected.

"As long as our **brain** is a **mystery**, the **Universe**, the reflection of the structure of the brain, **will also be a mystery.**"

Santiago Ramón y Cajal, 1920

CAMILLO **GOLGI**

Italian doctor Camillo Golgi invented a tissue-staining technique.

While studying the nervous system, Golgi (1843–1926) wanted to make the structure of nerve cells clearer under a microscope. In 1873, he discovered how to make them show up by staining them black with silver nitrate, a method Cajal greatly improved. Despite mounting evidence that he was wrong, Golgi clung tenaciously to the accepted reticular theory and even argued against Cajal's ideas when the two men collected their joint Nobel Prize.

Cajal used pen and ink to reproduce, with almost photographic precision, the intricate structure of the brain and nervous system. His drawings were a powerful tool for transmitting his observations to the scientific world.

German physicist Max Planck's quantum theory refuted the ideas of classical physics by showing that energy is emitted not continuously, but in fixed packets, or "quanta." It fundamentally altered the way scientists interpreted the subatomic world.

MILESTONES

HEAT THEORY WORK
Advances his study of heat theory from 1885 to 1889, while associate professor of physics at Kiel University.

STUDIES RADIATION
Investigates black body radiation and calculates the vibration frequencies of atoms during the 1890s.

MAKES BREAKTHROUGH
Presents his radiation distribution law in 1900, introducing the concept of energy quanta to physics.

HELPS EINSTEIN
In 1905, is one of the first prominent physicists to publicly support Einstein's Theory of Relativity.

AWARDED NOBEL PRIZE
Receives Nobel Prize in Physics in 1919 for his groundbreaking discovery of energy quanta.

Born in Kiel, northern Germany, Max Planck was the youngest of six children. When he was 9, the family moved to Munich, and Planck attended the Maximilian gymnasium. There, he showed an aptitude for mechanics, mathematics, and music, and after one tutor sparked his interest in physics, he chose to study the subject in college. At 21, he received a doctoral degree from the University of Munich, submitting a thesis on the second law of thermodynamics (how heat moves). In 1889, he became professor of theoretical physics at the University of Berlin, where he remained until he retired in 1926.

Black body radiation

In the 1890s, physicists were struggling to explain the absorption and emission of light and became preoccupied with "black body" radiation. In 1859, German physicist Gustav Kirchhoff had defined a black body as a hypothetical body that absorbs all the electromagnetic radiation, including light, that falls on it. When it is heated, a black body radiates energy in the form of electromagnetic waves with a range of wavelengths, including visible, ultraviolet, and infrared light.

However, during their experiments, physicists had noted that the wavelengths radiated by hot objects were not the same as those predicted by classical theories of thermodynamics. In 1894, while in Berlin, Planck turned his attention to black body radiation, beginning a quest to match theory with observation.

Planck popularized the work of his friend Albert Einstein in Germany and created a new professorship specifically for him in 1914, at the University of Berlin.

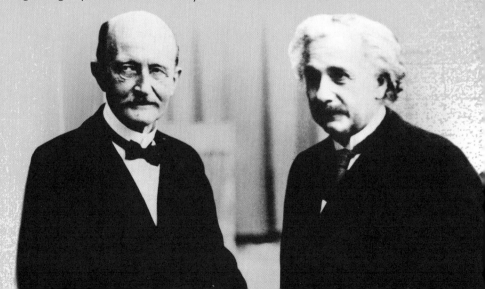

MAX **PLANCK**

1858–1947

"The **laws of Physics** have **no consideration** for the **human senses**; they depend on **the facts ...**"

Max Planck, 1931

Planck investigated how the intensity of the electromagnetic radiation emitted by a black body depended on the body's temperature and the frequency of the radiation (the color of the light). At the time, physicists assumed that the sources of radiation were atoms that could oscillate (vibrate) continuously at any frequency. But by 1899, Planck had noted that atoms could only vibrate at frequencies that were whole-number multiples of a base frequency that he named "h." For example, an atom could vibrate at 10h (because 10 is a whole number) but not 10.5h. Planck calculated the value of h, a quantity that is now called Planck's constant. It is a fundamental physical constant: its numerical value is the same everywhere in the known Universe. Planck also assumed that, contrary to the ideas postulated by classical physics, photons (particles of light) do

PAUL **DIRAC**

English physicist Paul Dirac is best known for his Dirac equation, which predicted the existence of antimatter particles such as the positron.

As a postgraduate student, Dirac (1902–1984) read Werner Heisenberg's paper on matrix mechanics describing how particles jump from one quantum state to another. He could see parallels with parts of the classical (prequantum) theory of particle motion. He worked out a way to understand classical systems on a quantum level and created quantum field theory. His Dirac equation predicted "antimatter" or "positrons"—particles with identical properties to particles of matter but with the opposite electrical charge. In 1932, Dirac was appointed Lucasian Professor of Mathematics at Cambridge University and, with Erwin Schrödinger, was awarded the Nobel Prize in Physics in 1933.

not emit energy in a smooth, continuous wave, but in measured amounts, or packets, later called "quanta" (from the Latin word *quantum*, meaning "how much"). A quantum is the smallest possible packet of energy.

Planck's radiation law

The theory, later called Planck's radiation law, finally explained the relationship between the temperature of an object and the energy emitted from that object in the form of electromagnetic radiation. Planck expressed this relationship as a mathematical equation: $E = h\nu$. The energy (E) in a photon equals its electromagnetic radiation frequency (ν) multiplied by Planck's constant (h).

In 1900, Planck presented his theory of light as "quantized" energy packets to the German Physical Society. Overturning all past physical theory, it helped initiate a revolution in physics, and today it is regarded as the origin of quantum theory. In 1905, Planck's hypothesis was verified by Albert Einstein, when he extended it in order to explain the photoelectric effect: the existence of discrete energy packets during the transmission of light.

Defining his constant h allowed Planck to devise a new set of physical units. Among these are the Planck length, the smallest unit of measurement possible: 1.6×10^{-35} meters. The amount of time it takes for a photon to travel a Planck length at the speed of light is one unit of Planck time. This is the smallest measurable unit of time: 5×10^{-43} seconds.

HIS **QUANTUM THEORY** IS **1** OF THE **2 FOUNDATION STONES** OF 20TH-CENTURY **SCIENCE**

NOMINATED FOR THE **NOBEL PRIZE** IN **PHYSICS 74** TIMES

THERE ARE **83** MAX PLANCK **INSTITUTES** WORLDWIDE

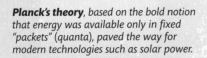

Planck's theory, based on the bold notion that energy was available only in fixed "packets" (quanta), paved the way for modern technologies such as solar power.

"THE INTRODUCTION OF QUANTUM THEORY LED NOT TO THE DESTRUCTION OF PHYSICS, BUT TO A SOMEWHAT PROFOUND RECONSTRUCTION."

Max Planck

The Universe in the Light of Modern Physics, 1931

Early-20th-century US biologist Nettie Stevens made the pioneering discovery that an animal's sex is determined by particular chromosomes. She was one of the first female scientists to be recognized for her contribution to genetics, although her work was largely overshadowed during her lifetime.

Nettie Stevens graduated from Stanford University with a masters in biology in 1900, before gaining a doctorate at Bryn Mawr College in 1903. Although her scientific career did not begin until she was 39, Stevens is credited today for her vital breakthrough in early genetics.

Sex determination

In the early 20th century, scientists were divided over how biological sex was determined. Many believed it was caused by external factors at the embryonic stage, such as temperature or nutrition. Stevens's research into the chromosomal behavior of mealworms ended this debate.

During her study, she noted that the male reproductive cells included X and Y chromosomes, but the females produced only Xs. She therefore concluded that the sex of an organism is determined by the chromosomes it inherits from each parent and published her findings in 1905. This research made the first link between a physical characteristic and a particular chromosome. However, Stevens's discovery was not widely acknowledged until after her death.

"Her work will be remembered."

Thomas Hunt Morgan, 1912

Stevens carried out painstaking work studying reproductive cells under a microscope, providing key data to support the theory of chromosomal inheritance. Her studies included many species of insects.

1861–1912

STEVENS

NETTIE

GEORGE
WASHINGTON
CARVER

African American agricultural scientist and experimenter George Washington Carver helped to restore the economy in the southern US via his innovative scientific methods of soil improvement and crop cultivation. He also developed commercially viable products that could be derived from crops grown instead of cotton.

MILESTONES

FURTHER EDUCATION
Becomes first African American to enroll at Iowa State Agricultural College; gains degree in 1894.

PRODUCT RESEARCH
From 1900 to 1920, invents 287 products derived from peanuts, as well as 118 from sweet potatoes.

RECOGNITION
Speaks for peanut farmers in front of the US House of Representatives in 1921 and receives a standing ovation.

MEDALS AND HONORS
Receives multiple honors, including the Spingarn Medal in 1923, recognizing outstanding achievement.

Born a slave on a Missouri plantation, Carver pursued an education following the abolition of slavery in 1865. After earning a master's degree in agriculture in 1896, he was made director of agriculture at the Tuskegee Institute, run by the renowned educator Booker T. Washington. Carver concurrently devoted himself to improving southern agriculture.

Restoring the economy
In the late 19th century, the main crop in the South was cotton, but exclusive cultivation of this crop had depleted nutrients from the soil and left yields at an all-time low. Carver encouraged farmers to grow peanuts, sweet potatoes, and soybeans instead, as these crops were nitrogen-rich and would help restore the soil. He then conducted research into derivative products that could be made from these crops, as the crops themselves were not commercially popular. Carver's work created over 400 marketable products, such as oils and dyes, which boosted financial yields and saw the South become a key contributor to the US agricultural industry.

"Education is the key to unlock the golden door of freedom."

George Washington Carver, 1896

Carver oversaw the Agriculture Department at Tuskegee Institute, Alabama, for 47 years. In his laboratory, he taught ex-slaves sustainable farming methods and techniques to become self-sufficient.

US geneticist and zoologist Thomas Hunt Morgan won a Nobel Prize for his groundbreaking work on heredity. Following the work of Gregor Mendel, he confirmed the role of chromosomes and laid the foundations for the modern field of genetics.

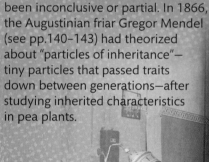

Morgan bred millions of fruit flies in his "Fly Room" laboratory. Able to produce a set of offspring in around 10 days, fruit flies were ideal for studying inheritance.

Thomas Hunt Morgan was born to a wealthy family in Kentucky in 1866. His countryside upbringing fueled a lifelong fascination with the natural world, from collecting fossils as a boy to conducting biological fieldwork during adolescence. Aged just 16, Morgan attended the University of Kentucky to study sciences before accepting a place to carry out postgraduate work in morphology and physiology at Johns Hopkins University. By 1890, aged 24, he had gained his PhD in zoology. In 1891, he became associate professor of biology at Bryn Mawr College, where he combined teaching with research into embryology. During his tenure at Bryn Mawr, Morgan made significant advances in experimental embryology, although it was to be his later work in genetics that earned him lasting fame.

The question of inheritance

In 1904, Morgan moved to New York to take up the role of professor of experimental zoology at Columbia University, and it was here that his groundbreaking work in heredity began. At the time, the field of genetics, as it is now known, did not exist—research had been conducted into inherited characteristics, but much of this had been inconclusive or partial. In 1866, the Augustinian friar Gregor Mendel (see pp.140–143) had theorized about "particles of inheritance"— tiny particles that passed traits down between generations—after studying inherited characteristics in pea plants.

THOMAS

HUNT MORGAN

1866–1945

Later scientists, including German biologist Theodor Boveri, had furthered this research, but the process behind inheritance remained unclear.

Early in his career, Morgan was skeptical about much of the previous research, including Mendel's theories of inheritance and Darwin's theory of natural selection. Naturally critical, and favoring controlled laboratory experiments over observational science, Morgan decided to carry out his own experiments to understand heredity.

The "Fly Room"

The subject Morgan chose for his experiments was *Drosophila*—fruit flies. They were ideal specimens: they displayed a wide range of physical traits, had only a small number of large chromosomes—which carry inheritable genetic information in the form of genes— and reproduce at great speed and in vast numbers. In 1908, Morgan set up a large laboratory at Columbia University, known as the "Fly Room," in which he

"We geneticists should rejoice, even with our noses to the grindstone."

Thomas Hunt Morgan, 1932

bred fruit flies by the millions. He cross-bred fruit flies with particular traits and analyzed the variations in their offspring by studying their chromosomes under a microscope.

Discovering inheritance

The key discovery came when a white-eyed male fly was bred with a red-eyed female. All the offspring had red eyes, indicating red was the dominant color while white was recessive. Morgan then cross-bred two of these offspring: one in four of the second-generation offspring had white eyes, and every

HERMANN JOSEPH **MULLER**

An eminent US geneticist, Hermann Joseph Muller received a Nobel Prize for his work on genetic mutations, which proved hereditary changes could be artificially induced through X-rays.

Brought up in New York, Muller (1890–1967) obtained a scholarship to Columbia University in 1907 to study biology and gained his PhD in 1916. He was inspired by the work of Thomas Hunt Morgan, and from 1920 spent 12 years at the University of Texas investigating genetic mutations. In 1926, he conducted a series of experiments that induced genetic mutations through the use of X-rays. This original and significant work secured his position as a renowned geneticist. Muller also became a pioneer in raising awareness of the long-term dangers of human exposure to radiation and campaigned for controls on nuclear weapons.

Morgan's experiments with fruit flies enabled him to trace the inherited physical characteristics across generations and determine the exact behavior of genes. The second generation of cross-bred flies revealed the white-eyed trait passed through the male line.

white-eyed fly was male. This showed that the white-eyed trait must be linked to the fly's sex. Further experiments revealed other sex-linked physical traits, and Morgan concluded that all these traits must be inherited together from the chromosome responsible for sex determination. Male sex chromosomes are X and Y, whereas the female sex chromosomes are both X, so Morgan could deduce that white eyes and other traits seen only in males must be carried on the Y sex chromosome. Further study revealed that genes occupy a specific location on a chromosome, and this enabled the production of genetic maps.

Through his fly experiments, Morgan confirmed the chromosomal theory of inheritance, and acknowledged the value and accuracy of Mendel's work. In extending Mendel's work on plants and applying it to animals, Morgan marked a turning point in the study of inheritance and launched the field of genetics.

STUDIED FRUIT FLIES FOR 17 YEARS

WROTE 370 SCIENTIFIC PAPERS

MARIE CURIE

A true pioneer, Marie Curie was a passionate and dedicated scientist whose work on radioactivity opened up a whole new world of knowledge. She discovered two new radioactive elements and laid the groundwork for the use of radiotherapy in the treatment of cancer.

Marie Curie was born Maria Skłodowska in Warsaw, Poland. Her life was hard from the start: by the age of 10, both her mother and her eldest sister had died. Although she was very bright, women in Poland at that time were not allowed to attend college, so Marie studied in secret while working as a governess. By the time she was 23, she had saved enough money to go to college at the Sorbonne, in Paris. There, she earned degrees in both physics and mathematical sciences, and also met and married the physicist Pierre Curie. The pair made a brilliant team.

Fascinated by the recent discovery by Henri Becquerel that some chemical elements were radioactive, Curie decided to pursue research in this field herself, and Pierre joined her. They worked tirelessly in the analysis of uranium ore (known as pitchblende) and, in 1898, identified two new radioactive elements within it: polonium (named after Curie's native Poland) and radium. The discovery earned them the 1903 Nobel Prize in Physics, which was awarded to the Curies and to Becquerel.

Fascinated by his wife's work, Pierre abandoned his own research to join Curie in hers. Together, they published a series of landmark joint papers.

MILESTONES

NEW ELEMENTS
Discovers two new radioactive elements with her husband in 1898: polonium and radium.

ATOMIC PHYSICS
Creates a new field of study, atomic physics, and coins the word "radioactivity" in 1898.

NOBEL PRIZE
Awarded the Nobel Prize in Physics for her and her husband's "extraordinary services" in 1903.

CANCER TREATMENT
Works with French chemist André-Louis Debierne to isolate radium in 1910, later used in radiotherapy.

WINS AGAIN
Makes history as the first person to be awarded two Nobels by winning the chemistry prize in 1911.

183

The French physicist Henri Becquerel was the first to discover natural radioactivity. In 1896, while investigating X-rays, he noticed that uranium emits radioactive particles.

Becquerel (1852–1908) was interested in phosphorescence (light-emitting substances). He wanted to see if there was any connection between phosphorescence and X-rays. He took a salt that contained uranium and was known to glow after exposure to light and put it on to some photographic plates, hoping that the uranium would absorb light and reemit it as X-rays. He left it in a dark drawer overnight, and the next day was amazed to find that the plates had been exposed. This proved that the uranium in the salt had emitted radiation. In recognition of his work, the unit of radioactivity is named after him: the becquerel.

husband's professorship of physics at the Sorbonne. She continued with her research and finally isolated pure radium in 1910. For her exceptional work on the extraction and properties of this element, she won a second Nobel Prize in 1911, this time for chemistry.

Danger and value of radioactivity
Although Curie had coined the term "radioactivity" to describe how energy is released when atoms disintegrate into a different form, she was unaware of the dangers of handling radioactive substances and carried samples and stored them in her desk. The penetrating power of the rays given off by uranium and other radioactive substances makes them potentially dangerous, and Curie started to show signs of radiation sickness. In spite of this, scientists—including Curie—were starting to realize

Then, in 1906, tragedy struck: Pierre was knocked down by a carriage and killed. Left with two young daughters to raise, Curie also managed to take over her

"One never notices what **has been done**; one can only see **what remains to be done.**"

Marie Curie, 1894

that, if used properly, radiation could have important applications in both medicine and science.

Working to save lives

The Curies believed that knowledge should be shared and never attempted to profit from their discoveries. During World War I, Curie used the money she received for her Nobel Prize to fund 20 mobile X-ray units, nicknamed Petites (little) Curies, to be sent to the battlefront to scan wounded soldiers. She even drove one of the trucks herself and trained others, including her daughter Irène, to take X-rays of the wounded. After the war, Curie helped to set up the Curie Foundation (now Institut Curie), which pioneered the first research into the treatment of cancer using radium. Curie toured the United States twice in order to raise money for this research.

Unfortunately, the detrimental effects of too much radiation meant that Curie's work was to prove fatal to her. She died of the bone marrow disease aplastic anemia at the age of 66, probably due to her prolonged exposure to radiation.

In 1894, Marie met Pierre in Paris. They would later marry and work collaboratively. Their joint research on the properties of radioactivity won them the Nobel Prize in 1903.

THE FIRST **WOMAN** TO WIN A **NOBEL PRIZE**

DONATED HER **NOBEL MEDALS** IN **WORLD WAR I**

ONE GRAM OF **RADIUM COST** $100,000 IN 1921

Researching the heart of matter, Ernest Rutherford transformed our understanding of the atom, identifying its components and revealing its inner structure. He showed that atoms can disintegrate into smaller constituents and described two types of radiation generated by the process: gamma and beta radiation.

Ernest Rutherford was born into a farming family in the village of Brightwater on the South Island of New Zealand. After winning a scholarship to Canterbury College (now the University of Canterbury), he gained three science degrees, then won an overseas scholarship and went to Cambridge University to work with the physicist J. J. Thomson. Impressed with the quality of Rutherford's research, in 1898, Thomson recommended him for a professorship in physics at McGill University in Montreal, Canada. Aged only 27, Rutherford was accepted at McGill, and set sail for Canada.

It was in Montreal that Rutherford conducted the work that won him the Nobel Prize—for his "investigations into the disintegration of the elements." Radioactivity had just been discovered by Henri Becquerel (see p.182), and soon afterward Thomson had identified the electron: a negatively charged subatomic particle. Up until then, atoms had been thought to be indivisible, but now it seemed that they were made up of smaller parts and that radioactivity was released as they decayed. Rutherford studied the radiation emitted by uranium and discovered that it was of two types, which he called alpha and beta radiation. He also found that alpha radiation was composed of positively charged particles and that

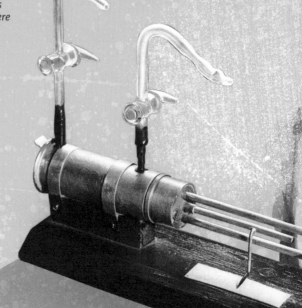

The first transmutation of the elements was achieved by Rutherford using this equipment (right). Nitrogen atoms were turned into oxygen by collision with alpha particles inside the tube. A window showed the protons emitted.

"I have broken the machine and **touched the ghost of matter.**"

Ernest Rutherford, 1917

ERNEST

RUTHERFORD

1871–1937

"The changes in question **are different ... from any** that have been **before dealt with in chemistry.**"

Ernest Rutherford, 1902

Rutherford fired *positively charged particles at a thin layer of atoms. Most of the particles went straight through the atoms, but some were deflected, and some bounced right back. He concluded that atoms consisted mainly of empty space with a tiny positively charged center, the nucleus, orbited by a cloud of electrons.*

NUCLEUS

CHARGED
PARTICLE

ELECTRON

when atoms release these particles, they become smaller. This led him to propose that atoms disintegrate and that radiation is a by-product of the process. Rutherford also noticed that different radioactive materials disintegrate at different rates, which he called their "half-lives," since the time it takes for each material to reduce by half could be predicted.

Inside the atom

When he returned to England in 1907, Rutherford became professor of physics at the University of Manchester—and it was there that he made his most famous discovery. At the time, the accepted model of the atom was one proposed by Thomson, Rutherford's former mentor. Thomson imagined that an atom was a diffuse cloud of positive charge in which electrons were embedded. This was known as the "plum pudding" model, and to test it, two of Rutherford's students fired alpha particles at an ultra-thin film of gold foil. They expected the particles to pass through the gold atoms, albeit with slight deflections. However, while some of the particles passed through, others rebounded, suggesting that there was something in their way. Rutherford concluded that the mass and positive charge of an atom were all concentrated into a very small volume at its center, which he called its "nucleus."

Radical shift

In 1911, Rutherford published his model of the atom, which he likened to a miniature solar system: mostly empty space, with electrons orbiting a tiny nucleus, all held together by energy. This idea marked a profound leap, as it suggested that matter, at its

NIELS **BOHR**

Along with Rutherford, Danish physicist Bohr played a key part in teasing out the structure of the atom.

Realizing that classical physics could not explain the Rutherford atomic model, Bohr (1885–1962) turned to quantum physics for the explanation. In 1913, he proposed that electrons have fixed levels of energy and orbit the nucleus at fixed distances in orbital "shells." Electrons change to higher or lower orbits by absorbing or emitting defined packets of energy (quanta).

most fundamental level, was not solid—a concept that was too radical for most of the scientific community, who liked the plum pudding model and thought that Rutherford's atom would not be able to remain stable.

Rutherford succeeded Thomson as Cavendish Professor of Experimental Physics at Cambridge University in 1919, where he continued his research. He found that if nitrogen and other light elements are bombarded with alpha particles, they emit hydrogen nuclei. Hydrogen was known to be the simplest element, and Rutherford surmised that its nucleus must be one of the building blocks of all the elements. The hydrogen nucleus had a positive charge, so he named it the proton. He later proposed the existence of particles that have a neutral charge and reside within an atom's nucleus alongside the protons. The neutron (the name for such particles) was finally identified in 1932, by James Chadwick at the Cavendish Laboratory, under Rutherford's guidance.

DEMONSTRATED **THAT ATOMS ARE NOT INDESTRUCTIBLE**

DISCOVERED THE **ATOMIC NUCLEUS AND THE PROTON**

HAD A SYNTHETIC ELEMENT NAMED AFTER HIM: **RUTHERFORDIUM**

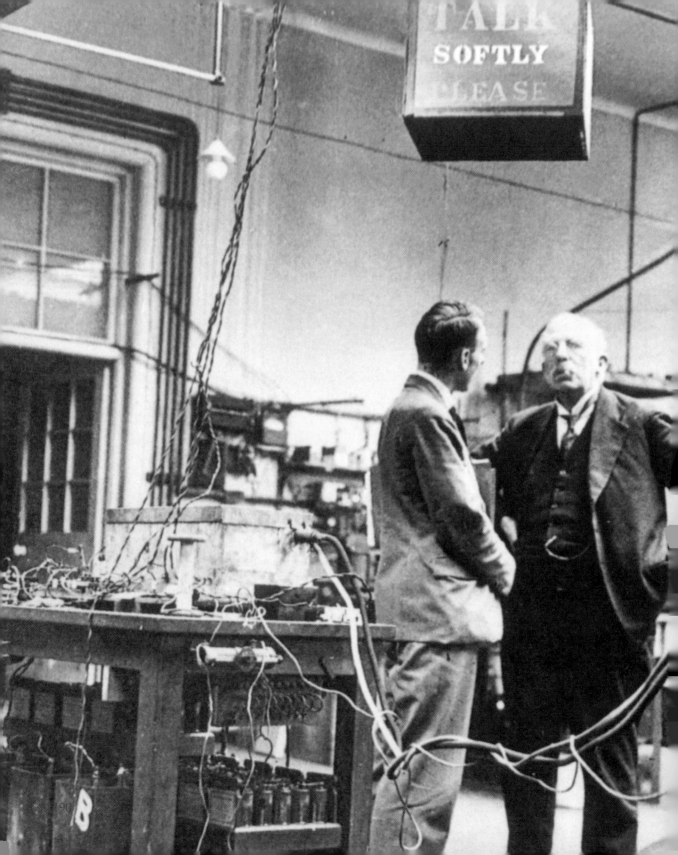

"I WAS BROUGHT UP TO LOOK AT THE ATOM AS A NICE HARD FELLOW, RED OR GRAY IN COLOR, ACCORDING TO TASTE. IN ORDER TO EXPLAIN THE FACTS, HOWEVER, THE ATOM CANNOT BE REGARDED AS A SPHERE OF MATERIAL."

Ernest Rutherford
At an after-dinner address given in 1934

◄ *Rutherford* **(right)** *photographed in 1934 in the Cavendish Laboratory at Cambridge University.*

ANTONIO EGAS

MONIZ

Highly regarded for his groundbreaking research in brain imaging, Portuguese neurologist António Egas Moniz also developed the controversial surgical operation that became known as the frontal lobotomy. His technique, refined by other neurologists, became a central part of psychiatric treatment in the 1940s and 1950s.

In 1927, António Egas Moniz, the University of Lisbon's first neurological professor, invented the angiogram, a type of X-ray for checking blood vessels. The invention was driven by Moniz's interest in how to better identify the position of brain tumors to improve the chances of their successful removal. He believed that if the blood vessels of the brain could be seen by radiography, tumors could be precisely located. Experimenting on human subjects with injections of radiopaque dyes, he found that a solution of 25 percent sodium iodine gave the safest and clearest result. Moniz's technique became the only diagnostic tool used to image cerebral vessels and identify blocked arteries until the advent of CT imaging in 1975.

Moniz was a pioneer in psychosurgery, too. He believed that he could eradicate certain mental states, such as obsession and depression, by operating on the frontal lobes of mentally ill patients. He designed a needle with a wire loop, a leucotome, for cutting through but not removing the connecting tissue of the lobes. His lobotomies (also known as leucotomies) on patients suffering from depression, anxiety, and schizophrenia were thought to alter associated personality traits and behaviors. The technique, further developed in the US, became a standard procedure for treating the mentally ill until the 1960s. However, only a third of patients benefited, and others experienced serious harm.

MILESTONES

VASCULAR IMAGES
Begins experimentation on vascular imaging with the help of Pedro Almeida in 1902.

ANGIOGRAPHIC PAPER
Publishes his first paper on angiography in the French journal *Revue Neurologique* in 1927.

NEAR-FATAL SHOOTING
Shot by a schizophrenic patient in 1939. His spine is shattered, but he continues to practice.

NOBEL AWARD
Receives Nobel Prize in Physiology or Medicine in 1949 for invention of prefrontal lobotomy.

Before the invention of brain scanners, angiograms—invented by Moniz—were one of the best ways to study abnormalities inside the brain.

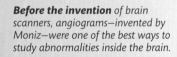

"His life was **unusually productive**; his name will live for his **two great contributions ...**"

Geoffrey Jefferson, 1955

Austrian theoretical nuclear physicist Lise Meitner coined the term "nuclear fission" and came up with the theory that explained the science behind the splitting of the nuclei of uranium atoms, first carried out by her colleagues in 1938.

MILESTONES

BERLIN RESEARCH
Moves to Berlin in 1907, working as departmental assistant at Max Planck's physics institute.

HEAD OF DEPARTMENT
Awarded Leibniz Medal and made supervisor of the physics section at the University of Berlin in 1917.

SETTLES IN SWEDEN
Escapes Nazi Germany in July 1938, when Austrian citizens become fully subject to German law.

REVOLUTIONARY PAPER
Along with nephew Otto Frisch, publishes theory of nuclear fission in *Nature* on February 11, 1939.

LONG-AWAITED AWARD
With Hahn and Strassman, given the 1966 Enrico Fermi Award by US Atomic Energy Commission.

Lise Meitner grew up in Vienna, the third of eight children in a liberal Jewish family. Her father Philipp was a lawyer who, together with his wife Hedwig, fostered an intellectually stimulating environment for their children. Meitner and her siblings were often present among regular gatherings of writers, lawyers, legislators, politicians, and chess players in the family home.

At the age of 8, Meitner kept a mathematics notebook under her pillow and asked probing questions of the world around her, including ones pertaining to reflected light and the patterned sheen on oil spills. Recognizing her intellect, and believing his daughters should enjoy the same education as his sons, Philipp ensured Meitner was privately tutored after she turned 14, the age at which girls in Austria were barred from state schooling.

Female trailblazer
The objective of Meitner's private tuition was to pass the entrance exam for the University of Vienna. She passed in July 1901, and later that year, aged 23, became one of the first women to attend the university's physics course. Having been tutored by the brilliant theoretical physicist Ludwig Boltzmann, in February 1906, Meitner became only the second woman at the university to receive a physics doctorate. Her thesis was on how heat travels through inhomogeneous solids (in which particles are not evenly distributed) and proved that they conduct heat in a similar way to electricity. This was in

Having fled to Sweden, Meitner was initially unable to see the results of her nuclear fission experiment. She explained the process of fission using data that Hahn sent by letter.

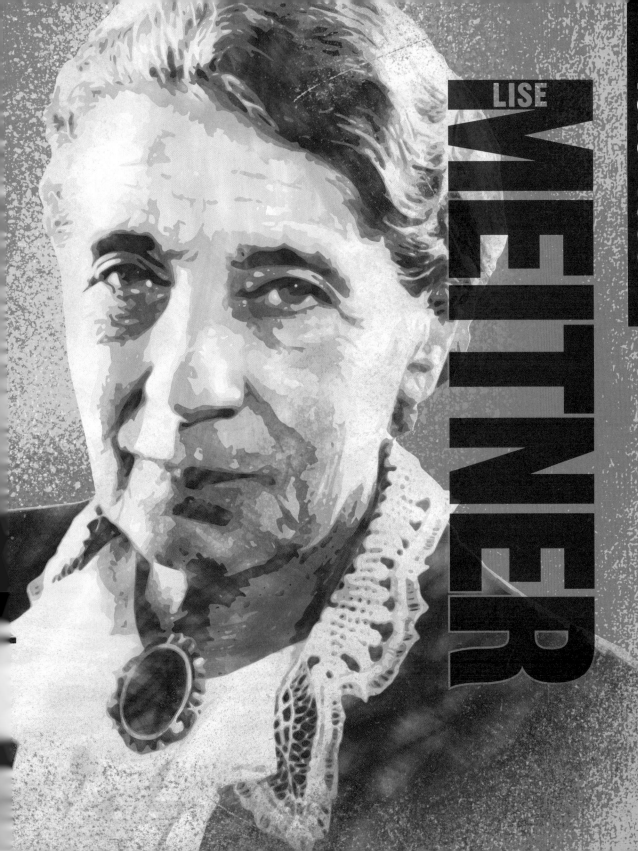

LISE MEITNER

1878–1968

NOMINATED FOR THE NOBEL PRIZE 48 TIMES BETWEEN 1924 AND 1965

BECAME THE FIRST FEMALE PHYSICS PROFESSOR IN GERMANY IN 1926

line with a formula developed by James Clerk Maxwell (see pp.148–151). Meitner spent much of the rest of 1906 absorbed in further research, studying among other things alpha and beta radiation.

Radiation research

In 1907, Meitner was invited by Max Planck (see pp.168–173), professor of physics at the University of Berlin, to conduct postdoctoral studies on radioactive substances. As was usual for female staff at the time, Meitner had no salary, but she worked with the leading minds of the day. Einstein (see pp.198–203) was one of her Berlin contemporaries, and she was soon introduced to Otto Hahn, a German radiochemist with whom she worked for the next 30 years. The duo studied the physical properties of radioactive elements and discovered a number of new isotopes (different forms of the same element). In 1913, Meitner was given the same salaried position as Hahn.

Between 1917 and 1918, Meitner, who had been serving as an X-ray technician during World War I, worked together with Hahn to discover a new isotope of the radioactive element protactinium. She went on to publish her findings on the Auger Effect, which causes an electron to be emitted from the outer shell of an atom, in 1922.

Although Meitner had converted to Christianity in 1908, the growing Nazi threat in the 1930s meant that her position in Germany grew more perilous. When Austria was annexed in 1938, she fled to Sweden, helped by the physicists Niels Bohr and Dirk Coster.

Nuclear discovery

Meitner took a post as a researcher at the Nobel Institute in Stockholm but remained in contact with Hahn in Berlin. On December 24, 1938, she received a report from Hahn of unexpected results in a study she had urged him to continue. When bombarded with neutrons, uranium appeared to "burst" and form barium, a much lighter element.

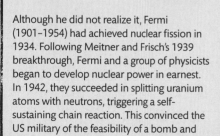

ENRICO **FERMI**

In 1942, Italian physicist Enrico Fermi created the first working nuclear chain reactor.

Although he did not realize it, Fermi (1901–1954) had achieved nuclear fission in 1934. Following Meitner and Frisch's 1939 breakthrough, Fermi and a group of physicists began to develop nuclear power in earnest. In 1942, they succeeded in splitting uranium atoms with neutrons, triggering a self-sustaining chain reaction. This convinced the US military of the feasibility of a bomb and the danger of Nazi Germany making one first.

NEUTRONS

Meitner discussed the matter with her nephew, Otto Frisch, a physicist in Denmark. Applying Einstein's theory of relativity, she realized that the mass was not lost, but was converted into energy. When the nuclei of the uranium atoms divided (nuclear fission), forming two new nuclei that together weighed less than the original uranium nucleus, an enormous amount of energy was created. Frisch repeated the results in his Copenhagen laboratory, and Meitner published their joint findings on the nuclear fission of uranium in early 1939. News of the discovery spread rapidly and ultimately led to the US military creating an atomic bomb (see box).

The Nobel Prize for the discovery of nuclear fission was awarded to Otto Hahn in 1944—Meitner's role went unrecognized for decades. However, she did not seem bitter about her omission, saying that "Science can bring both joy and satisfaction to your life."

"**Science** makes people reach selflessly for **truth and objectivity.**"

Lise Meitner, 1953

In Hahn's 1938 experiment, Meitner explained the process of nuclear fission, which occurs when a nucleus is bombarded with a neutron and then splits, releasing a vast amount of energy.

SPLIT NUCLEUS

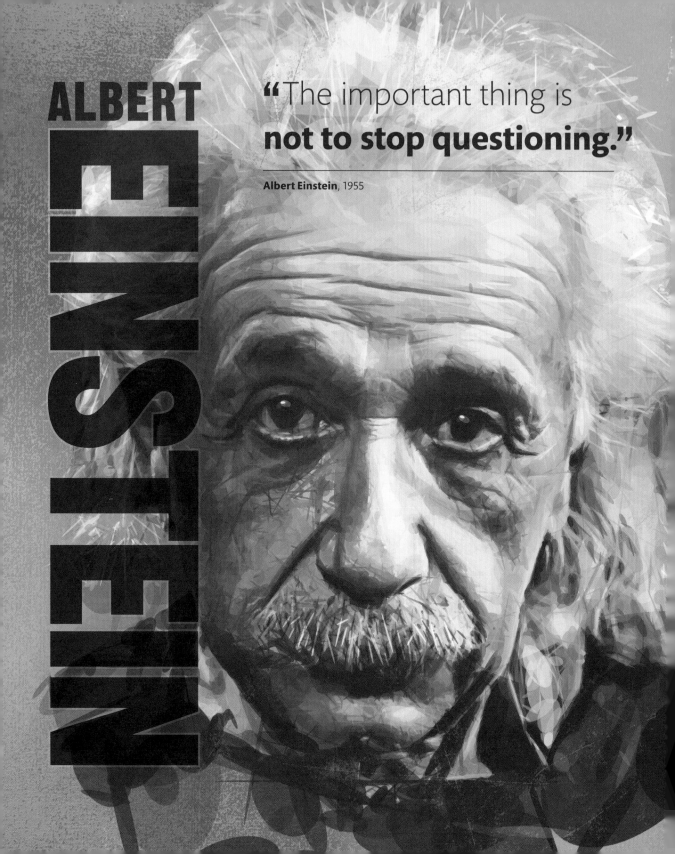

German-born physicist Albert Einstein developed the general and special theories of relativity, which utterly transformed the concepts of space, time, and gravity. His ideas were the most profoundly influential in 20th-century physics and paved the way for ground-breaking technologies, such as nuclear energy and solar power.

MILESTONES

PUBLISHES PAPERS
Writes four scientific papers in 1905 that bring attention from academics around the world.

SEMINAL WORK
While a professor at the University of Berlin, publishes "General Theory of Relativity" in 1915.

GLOBAL RECOGNITION
Awarded the Nobel Prize in Physics for services to theoretical physics in 1921.

LEAVES GERMANY
Emigrates to the US in 1933 due to the rise of Nazism and oppression of Jews in Germany.

ADVOCATES PEACE
Signs the Russell–Einstein Manifesto in 1955 to alert the public to the dangers of developing nuclear weapons.

Born to Jewish parents in Ulm in southern Germany, Albert Einstein had a natural talent for mathematics and a love of learning. He completed his schooling in Switzerland after spells in Munich and then Italy, where the family moved due to the failure of his father's electrical engineering business. After qualifying as a teacher, he was unable to secure a teaching post, and instead began work as a clerk at the Swiss patent office in 1902.

Einstein's relatively lowly position as a third-class technical expert gave him the necessary time to work on his own research interests. By considering the possible solutions to untestable problems (intuitive "thought experiments"), over the next few years, he made a series of breakthroughs that would transform the fundamental laws of physics.

Landmark findings

In 1905, Einstein submitted his doctoral thesis at the University of Zurich and had four landmark papers published in the journal *Annalen der Physik*. Two of these explained phenomena that had long puzzled scientists: the "photoelectric effect" and "Brownian motion." In elucidating

Einstein first became a university lecturer in 1908 at the University of Bern, Switzerland. His complex theories and eccentric persona earned him a dedicated following.

Born in Kolkata, India, S. N. Bose wrote of the behavior of photons—particles of light—in a 1924 paper sent to Einstein, who was so impressed that he had the paper published.

Bose's (1894–1974) formulations led Einstein to predict a fifth state of matter, in addition to the known states (solid, liquid, gas, and plasma). Einstein theorized that this fifth state—named the Bose–Einstein condensate (BEC)—would form when matter possessing a dense collection of very low-energy particles—named "bosons" after Bose—is super-cooled to temperatures close to absolute zero. Scientists first made BEC particles in 1995, heralding a new field in superfluids and superconductors.

the photoelectric effect, Einstein added to the foundations of quantum theory: the area of physics that deals with light, atoms, and subatomic particles. The second paper proved molecules and atoms exist through Brownian motion.

The third paper outlined his special theory of relativity, which explained the relative motion of objects moving at constant velocity. He showed that the speed of light is constant, however fast an observer is moving relative to the light's source. But time does not run at a single speed—clocks moving relative to

"Life is like riding a bicycle. To keep your balance, **you must keep moving.**"

Albert Einstein, 1930

ORIGINAL PATH

each other tick their seconds away at different rates, although the effect would only be observable at extremely high relative speeds. This led Einstein to conclude that space and time are intrinsically linked in a "continuum."

Energy equals mass

Einstein's fourth paper was an extension of special relativity that proved that mass and energy are relative and are aspects of the same property. He expressed this as $E = mc^2$, which states that the energy (E) of an object is the same as its mass (m) multiplied by the square of the speed of light (c^2). Since c^2 is a very large number—about 90 million billion—it follows that even a small amount of mass can contain a huge quantity of energy. This is the basis of nuclear power: atomic nuclei are broken apart and lose mass, resulting in large amounts of energy.

Over the next 10 years, Einstein developed relativity to include gravity. Though Isaac Newton had first described gravity, he had been unable to explain its origins or the laws of physics that govern it. In his 1915 general theory of relativity, Einstein explained that time slows down in intense gravitational fields, which cause time, matter, and light to bend as they pass close to very massive objects, such as the Sun.

Testing the theory

Einstein's predictions were put to the test in 1919, when physicists observing a total solar eclipse from several locations noted that stars appeared to be out of place due to the light they emitted being bent around the Sun. This shift made no sense under Newton's theory of gravity, but corresponded exactly with Einstein's general theory of relativity.

LEARNED **EUCLIDEAN GEOMETRY** WHEN HE **WAS 12**

PUBLISHED **OVER 300** SCIENTIFIC **PAPERS** IN HIS LIFETIME

Einstein proposed in his general theory of relativity that gravity, normally seen as a force, can also be understood as a distortion of space-time caused by objects with large mass, like a large ball lying on a rubber sheet that causes a smaller object to roll toward it.

DEFLECTED PATH

"IMAGINATION IS MORE IMPORTANT THAN KNOWLEDGE. FOR KNOWLEDGE IS LIMITED, WHEREAS IMAGINATION EMBRACES THE ENTIRE WORLD, STIMULATING PROGRESS, GIVING BIRTH TO EVOLUTION."

Albert Einstein
Cosmic Religion, 1931

◄ *After emigrating to the US*, Einstein found his permanent home in Princeton, New Jersey, where this photo was taken in 1940.

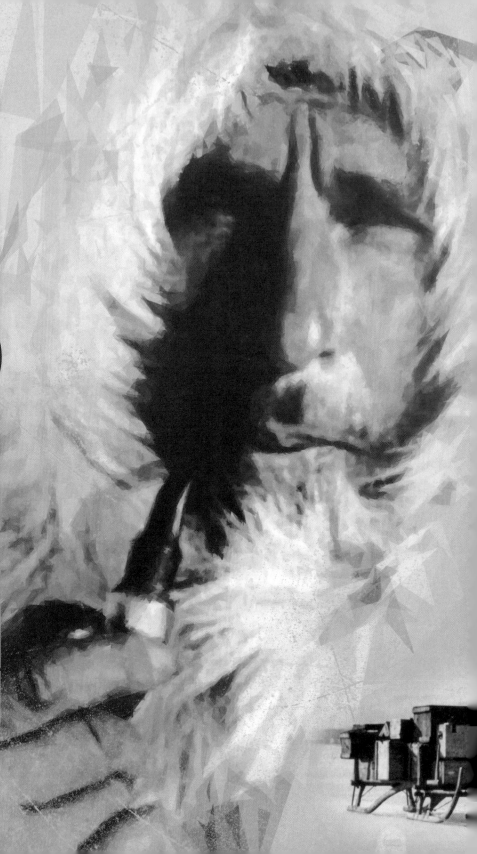

ALFRED WEGENER

A meteorologist and Arctic explorer, Alfred Wegener was the first to propose a systematic theory of continental drift. Although not trained in geology, he was an original thinker and was able to amass supporting evidence from his wide-ranging research. As with many innovators, his ideas were not accepted until long after his death.

Alfred Wegener was born in Berlin, Germany, the youngest of five children. He completed a PhD in astronomy at Berlin University, then switched to the field of meteorology (the study of weather and climate) and went to work at a meteorological station in the town of Beeskow. It was there that he, together with his brother Kurt, pioneered the use of weather balloons to study air movement.

In 1906, Wegener was given the job of meteorologist on a scientific expedition to chart the northeast coast of Greenland. It was an important learning experience, during which he built the first meteorological station in Greenland. On his return to Germany, he became professor of meteorology at Marburg University, where he began questioning why the continents are shaped the way they are. He had noticed that the coastlines

MILESTONES

FOSSIL RECORD
Discovers fossil evidence in 1911 that supports his theory that Africa and South America were once joined.

KEY PUBLICATIONS
Publishes his theory in two papers in 1912, and explains continental drift in a series of lectures.

CAREER INTERRUPTED
Drafted into the German army in World War I, but is released in 1914 after being wounded.

TRAILBLAZING BOOK
Publishes *The Origin of Continents and Oceans* in 1915, with maps of how the continents were joined.

ACADEMIC ROLE
Takes up the post of professor of meteorology and geophysics at Graz University, Austria, in 1924.

Wegener's fourth expedition to Greenland, in 1930, was to be his last. Just days after his 50th birthday, he set out across the ice for extra supplies and never returned. The expedition was completed by his brother.

"The theory is young and **treated with suspicion.**"

Alfred Wegener, 1915

of western Africa and eastern South America mirrored each other, like pieces of a jigsaw puzzle, and he wondered if they had once been joined together. In fact, he wondered if all of the Earth's continents had once been linked. He was not the first to notice the Africa–America pattern: it had been observed as early as the 16th century, and discoveries in the 19th century of the same types of fossils in both places had also attracted attention. Previous explanations included that the continents had been separated by the biblical Flood or that they were once joined by land bridges that had since sunk into the sea.

Cynognathus was a large doglike reptile that lived around 250 million years ago. Its fossils were found in both South Africa and South America.

Continental drift
Wegener dug deeper, searching in different scientific fields for more evidence. In the fossil record, he found further examples of the same species on different continents and discovered the same ancient rock formations on the African and South American coasts. He also found fragments of an ancient mountain range on unconnected continents and glacial deposits at the equator. All of this supported the idea that the continents had moved.

In 1912, Wegener presented his theory of continental drift, proposing that the continents had once been a single land mass, or "supercontinent," which he called Pangea. However, his idea was not well received. Geologists dismissed him for being an amateur

PRESENT DAY

CYNOGNATHUS

CYNOGNATHUS

CYNOGNATHUS

LAID THE **FOUNDATIONS** FOR THE THEORY OF

PLATE TECTONICS

CALCULATED THAT **ALL CONTINENTS** WERE JOINED AS A **SINGLE LAND MASS**

C.300 MILLION YEARS AGO

and pointed out that he could not explain how or why the continents had moved. To find more evidence, Wegener made three more expeditions to Greenland, but on the last, in 1930, he died from overexertion.

In the 1950s and 1960s, new evidence emerged that supported Wegener's theory. It was discovered that the Earth's crust is made up of gigantic tectonic plates, which move as convection currents bring molten lava up to the surface. Decades after Wegener's time, the jigsaw puzzle was solved.

MARIE **THARP**

Marie Tharp was a geologist and cartographer who charted the ocean floor. Her maps identified the mid-Atlantic rift—powerful evidence for the theory of continental drift.

During World War II, American women were encouraged to enroll in "masculine" disciplines, and Tharp (1920–2006) graduated in petroleum geology. She also learned technical drawing and got a job producing hand-drawn maps of the ocean depths from raw data. In 1953, she made a remarkable discovery—that there is a 9,941-mile (16,000-km) ridge in the middle of the Atlantic, with a deep rift at its center. Although she never argued for the theory of continental drift, her discovery supported it.

207

DIRECTORY

By the turn of the 19th century, science had become intimately linked with progress. As the 20th century unfolded, cutting-edge research fueled crucial shifts in fundamental ideas. Discoveries abounded and rules were rewritten at both subatomic and cosmological levels.

WILHELM CONRAD RONTGEN
1845–1923

German physicist Wilhelm Röntgen made one of the most important advances in physics and medicine when he discovered X-rays in 1895. He found that electrically charged vacuum tubes emitted rays that made a fluorescent screen glow. These electromagnetic rays went though human skin to expose photographic plates but were blocked by metal and bone. Although the discovery earned him the first Nobel Prize in Physics in 1901, he bequeathed the prize money to scientific research and never patented the X-ray. Röntgen is also known for his discoveries in mechanics, heat, and electricity.

IVAN PAVLOV
1849–1936

Russian-born Ivan Pavlov abandoned a religious career to become a professor at the Military Medical Academy in St. Petersburg in 1890. He was director of the physiology department at the Institute of Experimental Medicine when he began researching the digestive secretions of dogs. He found that the dogs learned to associate the arrival of food with the sound of a bell. After a while, even when no food was provided, the dogs still salivated in response to the bell being rung. This is now called classical, or Pavlovian, conditioning. In 1904, Pavlov won the Nobel Prize in Physiology or Medicine for his work.

KITASATO SHIBASABURO
1853–1931

Japanese physician and bacteriologist Kitasato Shibasaburō studied in Tokyo and Berlin and developed serum therapy to protect against tetanus and diphtheria. In 1890, he discovered that injections of his tetanus serum, which contained the antitoxin that had been produced in the blood of an animal exposed to the tetanus bacteria, conferred immunity on the animal to which it was given. He went on to apply the same principle to protect against diphtheria.

JULES HENRI POINCARE
1854–1912

Mathematical physicist Henri Poincaré was born in Nancy, France. While trying to prove that the solar system is stable, he noticed that even tiny changes in the initial conditions of a system often result in large—and unpredictable— changes in outcome; in other words, his results exhibited chaotic behavior. Published as early as 1908, his findings about chaos were initially overlooked, but several decades later, they became the foundation for chaos theory. He also wrote papers on electromagnetism that informed Einstein's work on relativity.

J. J. THOMSON
1856–1940

English physicist J. J. Thomson was one of the first scientists to describe the structure of atoms. He identified "corpuscles," later called electrons, using a cathode ray tube. The particles had a negative electric charge and were about 2,000 times lighter than a hydrogen atom. In his plum pudding model, Thomson suggested that every atom consists of a number of electrons and an amount of positive charge to balance their negative charges. He thought the positive charge was spread throughout the atom and the electrons existed within it, like plums in a plum pudding. His discovery revolutionized the theories of atoms and electricity. He also confirmed the existence of isotopes—elements that each have several types of atoms, chemically identical but differing in weight.

SVANTE ARRHENIUS
1859–1927

After studying physics at the University of Uppsala in his native Sweden, Svante Arrhenius became professor of physics at the University of Stockholm. In the 1890s, he decided that past ice ages might have been caused by fewer volcanic eruptions pumping gases such as carbon dioxide into the atmosphere. These gases retain heat, so reducing them would, he argued, cool down Earth. He also noted that burning fossil fuels would increase these gases and make the Earth warm up; in this way, he linked human activity with rising global temperatures, paving the way for modern concerns about climate change.

ANNIE JUMP CANNON
1863–1941

American astronomer Annie Jump Cannon was the 20th century's leading authority on the spectra of stars. Born in Delaware, she studied physics and astronomy at Wellesley College and joined the Harvard College Observatory in 1896 as one of several women hired to process astronomical data. She pioneered the classification of stars with her 1901 system, which laid the foundations for the Harvard Spectral Classification system. Over 44 years, she classified 350,000 stars.

HENRIETTA SWAN LEAVITT
1868–1921

While studying at Radcliffe College in Cambridge, Massachusetts, Henrietta Swan Leavitt became interested in astronomy. She went on to examine the luminosity of stars from thousands of photographic plates at the Harvard College Observatory. She saw that Cepheid variable stars (pulsating stars) showed a regular pattern of brightness. Her work was crucial for measuring the distance between the Earth and other galaxies. Leavitt discovered more than 2,400 variable stars and four novae (bright, transient interactions between

stars). She also developed a standard for photographic measurements, now called the Harvard Standard. Her work was largely unrecognized in her lifetime.

HARRIET BROOKS
1876–1933

Born in Ontario, Canada, Harriet Brooks graduated from McGill University in 1901 and became Canada's first female nuclear physicist. She studied under J. J. Thomson in Cambridge, UK, and Ernest Rutherford in Canada and worked in Marie Curie's laboratory in Paris from 1906. She discovered that one element could change into another through nuclear decay. Several of her findings were only attributed to her after her death.

SRINIVASA RAMANUJAN
1887–1920

Born in Madras (now Tamil Nadu), India, Srinivasa Ramanujan made major contributions to mathematical analysis and number theory despite little formal training. He was invited to Cambridge in 1913, after sending mathematician G. H. Hardy a letter and 120 mathematical theorems. Ramanujan was awarded a Bachelor of Science degree in 1916, and soon after became the second Indian to be elected a Fellow of the Royal Society. He contracted tuberculosis in 1916 and returned to India 2 years later, but died in 1920.

ERWIN SCHRODINGER
1887–1961

Born in Vienna, Austria, in 1887, Erwin Schrödinger studied physics at the University of Vienna. He moved to Germany, then to the University of Zurich, Switzerland, where he carried

out his most important work, immersing himself in the emerging field of quantum physics. Schrödinger's wave equation allowed a new understanding of how some particles behave. Rather than imagining electrons as orbiting the atom's nucleus, arranged in shells and subshells, the wave equation shows that orbitals, shells, and subshells are actually "clouds" of probability that tell us how likely it is for a particular electron to be found in a specific position. Schrödinger's equations of wave mechanics changed the way we see the world and formed the basis for today's quantum mechanics.

RONALD FISHER
1890–1962

British statistician and geneticist Ronald Fisher pioneered the application of statistics to scientific experimentation. In 1918, he published a paper that illustrated the use of statistical tools to reconcile what were apparent inconsistencies between Charles Darwin's ideas of natural selection and the recently rediscovered experiments of botanist Gregor Mendel. Fisher was knighted in 1952.

HAROLD UREY AND STANLEY MILLER
1893–1981; 1930–2007

In 1953, American chemists Harold Urey and Stanley Miller simulated early Earth in the laboratory with electrical sparks to imitate lightning. They used a closed series of connected glass flasks, sealed from the atmosphere, and placed in them water and a mixture of gases thought to have been present in Earth's primitive atmosphere—hydrogen, methane, and ammonia. They showed that with enough heat and energy, simple life-giving, carbon-based compounds could be produced.

6

WAR AND MODERNITY

1925–1950

British bacteriologist Alexander Fleming's accidental discovery of penicillin—the first antibiotic—was one of the greatest breakthroughs in the fight against disease. But it took another decade before other scientists managed to transform his research into a miracle drug that would save millions of lives.

MILESTONES

CHANCE FINDING
Discovers *Penicillium notatum* by accident in 1928, in an abandoned Petri dish in his laboratory.

KEY PUBLICATION
Publishes his findings in the *British Journal of Experimental Pathology* in 1929.

FIRST TREATMENT
Penicillin is trialed for the first time in 1941, on Albert Alexander, but he dies due to insufficient supplies.

WAR WOUNDED
Between 1943 and 1945, hundreds of thousands of war casualties are treated with penicillin.

JOINT AWARD
Awarded Nobel Prize in Physiology or Medicine jointly with Florey and Chain in 1945.

Penicillin was available as a treatment to many wounded soldiers during World War II. In 1944, it was used on troops in the D-Day landings, and had a huge impact on reducing the death toll.

Alexander Fleming was born into a poor farming family in Ayrshire, Scotland. At 16, he went to work for a shipping company, but 4 years later, he inherited some money from an uncle and decided to use it to study medicine. In 1906, he qualified with distinction at St. Mary's Hospital Medical School, London.

When World War I broke out, Fleming joined the Royal Army Medical Corps. Based at a military hospital in France, seeing first-hand the widespread deaths of soldiers from infected wounds deeply affected him. He realized that the antiseptics used were actually causing harm when applied to deep wounds. When he returned to his research post at St. Mary's at the end of the war, he resolved to study "bacteria killers." In 1923, he identified the natural enzyme lysozyme, which has antibacterial properties, but the research was of limited success.

"I did not invent penicillin. Nature did that."

Alexander Fleming

ALEXANDER

FLEMING

1881–1955

On September 3, 1928, a chance discovery changed everything. Fleming had been on vacation for weeks, and on his return, he started to clean up his notoriously messy laboratory. He was studying the bacterium *Staphylococcus*, the strains of which cause many diseases. Petri dishes containing the bacterium were stacked in the sink, waiting to be cleaned.

An accidental find

Fleming noticed something surprising. One of the dishes had been infected by some sort of mold, and the area around it was free of bacteria. Fleming grew samples of the mold, identified it as a type of *Penicillium*, and confirmed that it was killing the bacteria by secreting a substance—an antibiotic—which he called penicillin. This was not the first time a scientist had noticed that mold

"For the birth of something new, there has to be a happening."

Alexander Fleming

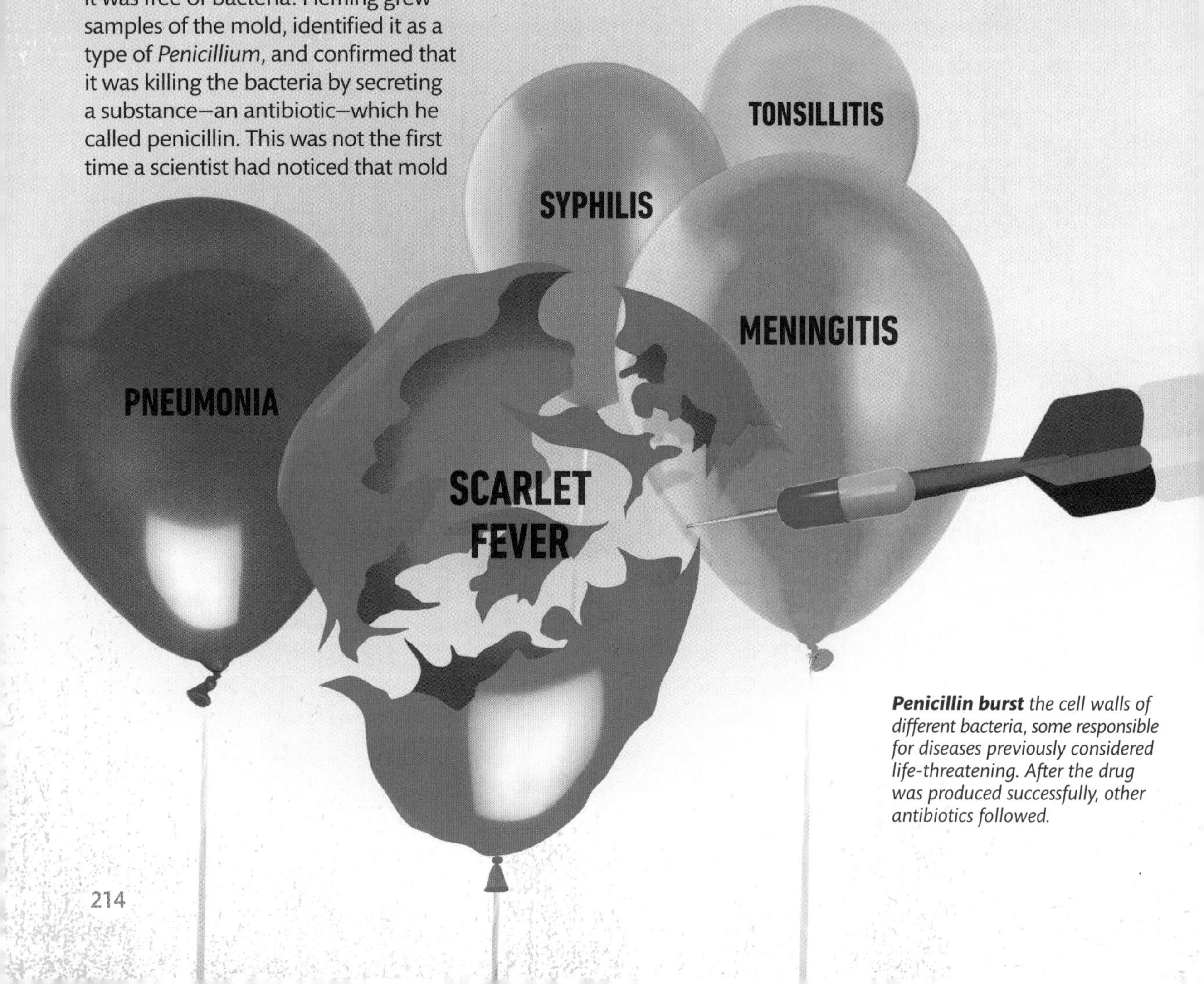

TONSILLITIS

SYPHILIS

MENINGITIS

PNEUMONIA

SCARLET FEVER

Penicillin burst the cell walls of different bacteria, some responsible for diseases previously considered life-threatening. After the drug was produced successfully, other antibiotics followed.

had antibacterial properties. Indeed, traditional remedies for infection had included moldy bread or fruit. When Fleming published his findings in 1929, initial prospects for exploiting the antibiotic were mixed: the mold was difficult to cultivate, the quantities of penicillin produced were tiny, and it could not be stored for long. As a result, Fleming effectively lost interest and gave up on the project for the next decade.

Fortunately, in 1939, there was another breakthrough: a team at Oxford University headed by Howard Florey (see box) came across Fleming's research and were able to radically improve the cultivation, extraction, and purification processes for penicillin. In 1941, the antibiotic was trialed on a man named Albert Alexander, who had abscesses and blood poisoning. Injections of penicillin had a dramatic effect, but the supplies

ran out, and sadly Albert died. However, four of the next five patients recovered from their infections. Drug companies became interested and invested in developing higher-yielding strains of *Penicillium*. Finally, penicillin could be made on a large scale. It proved highly effective against a wide range of bacterial infections, including scarlet fever, meningitis, and pneumonia.

What had once been unimaginable was now possible: bacterial infection could be effectively treated. The landscape of medicine had changed.

HOWARD **FLOREY**

Australian Florey led the team who learned how to produce penicillin in large quantities.

Florey (1898–1968), a pathologist at Oxford University, was researching natural antibacterials with biochemist Ernst Chain. Aided by Norman Heatley, they developed the expertise to manufacture and purify penicillin. Using all kinds of equipment, and with funding from the US, they made sufficient quantities of penicillin to test it first on mice and then on a human. Fleming and Florey were both knighted for their work.

IN **WORLD WAR II,**
TWO-THIRDS
OF **SOLDIERS** WITH **CHEST WOUNDS** WERE **SAVED BY PENICILLIN**

10 DOSES
OF PENICILLIN WERE MADE IN 1942;
600 BILLION
DOSES IN **1945**

The mathematician Emmy Noether had a brilliantly creative mind. Her ability to make conceptual connections in a whole new way transformed the fields of abstract algebra and theoretical physics.

Amalie Emmy Noether was born in the German town of Erlangen. She gained a doctorate in mathematics but struggled with discrimination against women in academia. Working for years without pay, she had to teach under the name of a male colleague. Despite this, she succeeded in winning admiration for her work. She is best known for Noether's theorem, which connects two key concepts within physics: conservation laws and symmetries. Each law can be associated with a particular symmetry and vice versa. Noether was described by Albert Einstein as a "creative mathematical genius."

MILESTONES

FEMALE GRADUATE
Gains a doctorate in mathematics in 1907: only the second woman to receive that degree.

LANDMARK THEOREM
Proves her theorem in 1918, fundamental to both particle physics and general relativity.

GAINS RECOGNITION
Awarded the highly respected Ackermann-Teubner Memorial Prize in mathematics in 1932.

FLEES GERMANY
Forced by the Nazis to leave Germany in 1933. Moves to the United States and takes up a post there.

> **"My methods** are really methods of **working and thinking."**
>
> **Emmy Noether**, 1931

EMMY NOETHER

INGE LEHMANN

Seismologist Inge Lehmann's precise data analysis revealed a secret at the heart of the Earth—its solid core.

Born in Copenhagen, Denmark, Inge Lehmann gained a degree in mathematics and went on to study earthquakes. Scientists knew that shock waves from an earthquake—P, or primary, waves and S, or secondary, waves—traveled through the Earth and could be detected on the other side of the planet. They had also observed shadow zones, where few waves were detected, and thought they occurred because the shock waves were deflected by the Earth's core, which was liquid. However, this did not explain the occurrence of weak P waves in the shadow zone. In 1936, Lehmann proposed that these P waves were deflected by the boundary between the inner and outer regions of the Earth's core and that the inner region was solid and not liquid as previously thought.

"The master of a black art."

Bertha Swirles, 1994

Lehmann discovered that the core of the Earth is more intriguing than was thought because the inner core rotates separately.

Indian physicist Sir Chandrasekhara Venkata Raman was awarded the 1930 Nobel Prize in Physics for his work on the scattering of light and for the discovery of the effect named after him. He also promoted the growth of science in India.

MILESTONES

TEENAGE PRODIGY
Gains a degree from the University of Madras in 1904, earning his master's degree 3 years later.

NEW DIRECTION
Leaves accountancy to become professor of physics at the University of Calcutta in 1917.

PROMOTES SCIENCE
Sets up the *Indian Journal of Physics* in 1926; founds the Indian Academy of Sciences in 1934.

LIGHT EXPERIMENT
In 1928, publishes leading research into the behavior of light particles; this wins him a Nobel Prize in 1930.

RESEARCH INSTITUTE
Becomes the founder and first director of the Raman Research Institute in 1948, based in Bangalore.

Born in Madras State, now Tamil Nadu, in southern India, C. V. Raman was just 14 when he enrolled at the University of Madras. After graduating, he went on to gain a master's degree. In 1907, he became an accountant but continued with scientific research in his spare time. Raman established a reputation as a highly able physicist and in 1917 was made professor of physics at the University of Calcutta, where he conducted extensive research into acoustics and optics.

Seeing the light
Raman's voyage to London in 1921 inspired him to investigate why the sea appears blue. He discovered that when a beam of light shines on a transparent substance, some of that light is scattered and a small amount alters in wavelength. This prompted his extensive series of experiments to measure how light was scattered by liquids, solids, and gases and led to his discovery of the Raman effect. This effect occurs when photons (the fundamental particle of light) either take up energy from or lose energy to a molecule. They then scatter with increased or decreased energy and a different wavelength. This explained why water molecules in the sea scatter white sunlight and the resulting light falls mainly in the blue part of the light spectrum. The Raman effect was considered by many to confirm Einstein's quantum theory, and it remains a key technique in substance analysis. In 1930, Raman became the first Asian to be awarded a Nobel Prize. He devoted the rest of his career to teaching students, founding academic publications, and setting up research institutions.

Raman, pictured here at his Raman Research Institute, discovered the Raman effect— which is now the basis of a laboratory technique called Raman spectroscopy.

"Science ... is a fusion of [one's] **aesthetic** and **intellectual functions** devoted to the **representations** of **nature.**"

C. V. Raman

C. V. RAMAN

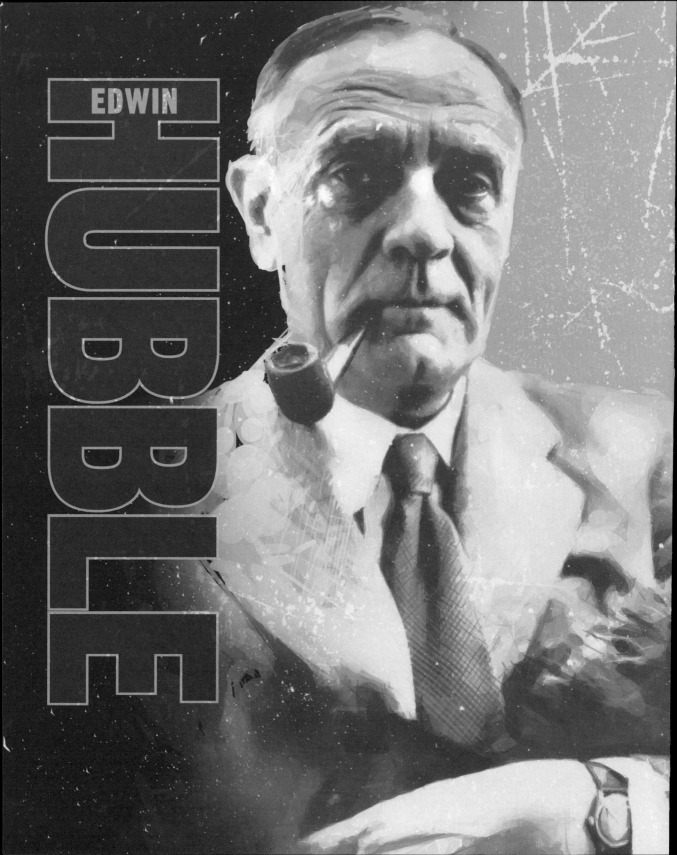

EDWIN
HUBBLE

US astronomer Edwin Hubble's meticulous research in the early 20th century redefined our understanding of the cosmos. He proved that the Universe is made up of millions of galaxies, not just the Milky Way (our galaxy), and that it is constantly expanding. These discoveries form the basis of modern cosmology.

Born in Missouri, Edwin Powell Hubble studied mathematics and astronomy at the University of Chicago before relocating to England to read philosophy of law at the University of Oxford to please his father. Returning to the US in 1913, he decided to further his studies in astronomy and, in 1917, received a PhD from Yerkes Observatory. Two years later, he joined the Mount Wilson Observatory in California—home of the Hooker telescope, the word's largest astronomical instrument—as a research astronomer, and spent his entire career there.

In the early 1920s, astronomers were divided into two schools of thought concerning the composition of the Universe: one school believed that the Milky Way was the entire Universe; the other, that it was just one of many galaxies in the Universe. It was Hubble who conclusively proved which theory was correct. Central to his work was the question of whether spiral nebulae (now known as spiral galaxies) lay beyond the Milky Way. In 1923,

MILESTONES

FIRST DISCOVERY
Observes that the Universe is bigger than previously believed while using the Hooker telescope in 1924.

CLASSIFIES GALAXIES
Devises the Hubble Sequence in 1926, which categorizes galaxies by shape, light, and distance.

PUBLISHES KEY PAPER
Links redshift of a galaxy to its distance from Earth in 1929: first evidence that the Universe is expanding.

EINSTEIN VISITS
Meets Einstein at the Mount Wilson Observatory in 1931; shares his evidence of an expanding Universe.

> "Man **explores the universe** around him and calls the **adventure Science.**"

Edwin Hubble, 1929

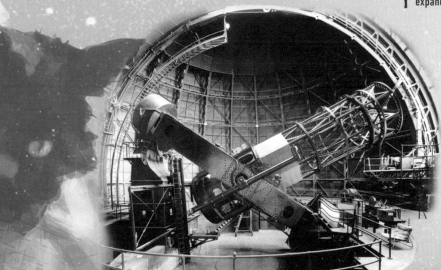

Hubble photographed *the nebulae in deep space using the powerful Hooker telescope and was the first to realize they were not gas clouds or distant stars in the Milky Way, as was believed, but galaxies.*

221

Hubble revealed that galaxies are moving away from Earth with speeds proportional to their distance. The result of this motion is that as time passes, the galaxies themselves are also moving farther apart, meaning that the Universe is expanding.

> " The **history of astronomy** is a history of **receding horizons.** "

Edwin Hubble, 1936

GEORGES **LEMAITRE**

Belgian astronomer, cosmologist, and Roman Catholic priest Georges Lemaître was the first scientist to propose the theories of an expanding Universe and the Big Bang.

In 1927, Lemaître (1894–1966) published a paper hypothesizing that the Universe is expanding. While his theory was largely overlooked at the time, today scientists agree that Lemaître should share credit for the discovery with Hubble, whose findings were published 2 years later. In 1931, Lemaître proposed that the Universe began as a single dense particle from which all energy and matter derived. This idea formed the basis of the Big Bang theory, evidence for which was discovered in 1964.

Hubble used the Hooker telescope to observe Cepheid variable stars (stars whose luminosity pulsates) within the spiral nebula Andromeda. By plotting the Cepheids' periods (the time for one cycle) of variation, he established the stars' absolute magnitude (full luminosity) and then compared this with their apparent magnitude (brightness from Earth) to calculate how far away they were. The results revealed that Andromeda is a vast, independent galaxy that is comparable in size to the Milky Way and located far beyond it—almost

1 million light-years away. This proved beyond doubt that the Universe was far larger than had previously been thought.

Investigating redshift

Redshift is a key concept in astronomy that was discovered by the American astronomer Vesto Slipher. He reported that spiral nebulae travel away from Earth after observing that the light they emit shifted toward the red end of the electromagnetic spectrum (rather than the blue end, which would indicate movement toward Earth). Hubble also examined the relationship between the distance of a galaxy from Earth and its redshift (when the light it emits increases in wavelength). He built on the research of other astronomers and verified Slipher's findings. He collected new data from other galaxies, which he plotted on a graph showing their distances relative to their redshifts.

Expanding Universe

Hubble's graph revealed a linear pattern, which indicated that the more distant the galaxy, the greater its redshift and therefore the faster it was receding from Earth. This phenomenon became known as Hubble's law. Hubble's surprising results, published in a landmark paper in 1929, made it clear that the existing notion of a stable and stationary Universe was incorrect and that we live in a large, dynamic Universe that is continually expanding in all directions. Since Hubble's discovery, scientists have drawn on his work to calculate the size of the observable Universe, to determine its age, and to estimate the moment of its origin.

STUDIED THE **REDSHIFTS** OF

46

GALAXIES

CALCULATED **ANDROMEDA** TO BE 900,000 **LIGHT-YEARS AWAY**

ESTIMATED **THE UNIVERSE** WAS AROUND **2 BILLION** YEARS OLD

FREDERICK BANTING

MILESTONES

BECOMES DOCTOR
Switches from divinity to medicine in 1916; receives medical degree from University of Toronto.

MILITARY HERO
Serves in the Canadian Army Medical Corps and is awarded the Military Cross in 1919.

DIABETES RESEARCH
In 1920, explains his idea for isolating insulin to Professor John Macleod at University of Toronto.

NOBEL PRIZE WINNER
Awarded the Nobel Prize in 1923 and shares the money with his assistant Charles Best.

The hormone insulin, isolated by Canadian physician Sir Frederick Banting in 1921, fast became the standard treatment for diabetes.

Frederick Grant Banting studied medicine at the University of Toronto. At the time, reports suggested that diabetes was caused by a lack of insulin—a hormone that controls the metabolism of sugar—but no one knew how to extract it from the pancreas. The trouble was that other digestive enzymes destroyed it. In 1920, Banting read that closing the pancreatic duct kills the acini cells of the pancreas, which make digestive enzymes, but leave the cells that secrete insulin intact.

Banting began by operating on dogs, closing their pancreatic ducts before operating again to extract their insulin. He then injected this into diabetic dogs, and it successfully lowered their sugar levels. In 1922, 14-year-old Leonard Thompson became the first human being to receive this treatment—for which Banting received the Nobel Prize in 1923.

Banting loved dogs, and it upset him to use them in his insulin trials.

" The **greatest** **joy** in life is to **accomplish** ..."

Frederick Banting, 1928

US chemist Alice Ball created an injectable form of chaulmoogra oil extract. Known as "the Ball Method," it became the key treatment for the disfiguring disease leprosy until the 1940s.

Born in Seattle, Alice Augusta Ball studied chemistry at the University of Washington. She moved to Hawaii, and in 1915, became the first woman and the first African American to graduate from the College of Hawaii (now the University of Hawaii) with a master's degree in science. While teaching there, she researched a way to make injected chaulmoogra oil tolerable. Doctors knew the oil could treat leprosy, but when injected, it clumped under the skin, forming blisters. Ball exposed the oil's fatty acids to a catalyst to produce water-soluble ethyl esters (chemical compounds). When injected, the oil then dispersed. By 1922, her technique was widely used.

Ball died in 1916, a year after her discovery, and her college president was credited with the technique. In 2007, she was finally honored by the University of Hawaii with its Medal of Distinction.

MILESTONES

MASTER'S FIRST
Becomes the first woman and first African American to receive a master's degree in 1915.

BALL METHOD
Isolates chaulmoogra oil chemicals in 1916 to create an injectable form, which treats leprosy.

UNTIMELY DEATH
Dies in 1916, aged just 24, before her treatment is rolled out around the world.

WINS RECOGNITION
In 2000, 84 years after her death, the governor of Hawaii declares February 29 "Alice Ball Day."

"**Who knows** what other **marvelous work** she could have **accomplished.**"

Paul Wermager, 2017

ALICE BALL

WALLACE

CAROTHERS

US chemist Wallace Carothers led the research team that invented neoprene, the world's first synthetic rubber, and nylon, the first fully synthetic fiber. His work altered the course of organic chemistry and ushered in the modern age of cheap plastics and other synthetic materials.

Born in Iowa, Wallace Hume Carothers was a chemistry instructor at Harvard University when, in 1928, the DuPont chemical company recruited him to investigate the chemical modification of natural materials to produce artificial ones. He studied polymers—substances found in all living organisms whose structure and chemical properties were not fully known. By 1929, Carothers believed that polymers were very large molecules made when smaller molecules join to form a long, repeating chain.

His idea was validated in the laboratory in 1930, when a test tube was found to contain a polymer similar to isoprene, found in natural rubber. The team had made a synthetic rubber, later dubbed neoprene. DuPont then tasked Carothers with finding a synthetic replacement for silk. In 1934, after experimenting with hundreds of chemical compositions, he found a formula for a super-strong, elastic polymer that could be drawn out into thin threads and woven into fine fabric: DuPont patented it as "nylon."

"The best organic chemist in the country."

Roger Adams

MILESTONES

JOINS DUPONT
Appointed director of research in organic chemistry in 1928, with brief to research polymers.

PUBLISHES RESEARCH
Writes articles on polymer theories in 1929 and 1931, establishing foundation of modern polymer science.

DISCOVERS NEOPRENE
Accidentally discovers artificial rubber with his team in 1930, using the polymerization process.

MAKES MIRACLE FIBER
Creates first human-made fiber, nylon, in 1934. Its first commercial use is in toothbrush bristles.

In May 1940, nylon stockings first went on sale in the US, and 800,000 pairs were bought. Demand continued unabated, and the name "nylons" became another word for stockings.

227

PERCY JULIAN

African American Percy Julian overcame racial barriers to be one of the greatest research chemists of the 20th century.

As a research fellow at DePauw University, Percy Julian was the first to chemically produce the anti-glaucoma drug physostigmine, making it widely available. Racial prejudice forced him out of DePauw, and he moved to a soybean oil company. Here, he pioneered the large-scale synthesis of human hormones from plants and became a world-class research scientist.

Mass drug production

In 1940, Julian devised a cost-effective way to extract the steroid stigmasterol from soybean oil and convert it into the hormone progesterone on an industrial scale. His work led to the synthesis of many other hormones from natural sources and helped to secure their widespread use.

MILESTONES

CHEMISTRY PhD
Graduates from DePauw as valedictorian in 1920; gains PhD in chemistry in 1931 from University of Vienna.

DRUG SYNTHESIS
Achieves a total synthesis of physostigmine in 1935; publishes results in a landmark research paper.

HUMAN HORMONES
Leads key research into the synthesis of human hormones from plant products in the 1940s.

PLANT STEROIDS
Sets up own company, Julian Laboratories, in 1953, continuing in the synthesis of plant steroids.

NATIONAL ACCLAIM
Becomes the first African American chemist elected to the National Academy of Sciences in 1973.

> "We **dared** to do something rather **bold.**"

Percy Julian, 1965

The discovery of stigmasterol as a by-product of soybean oil resulted from an accident. Water leaked into a tank of oil, creating a white solid: Julian identified it as the plant-derived steroid.

Leading US geneticist Barbara McClintock's innovative discoveries changed our understanding of genes and their behavior.

Barbara McClintock studied plant genetics, specifically the chromosomes of corn. In 1931, she wrote a paper in which she proposed that chromosomes were the bedrock of genetics, and established herself as the leader in the field. She later discovered that genetic information is not static but can move within or between chromosomes—a phenomenon known as "jumping genes." Her studies also showed that certain genetic elements can influence the behavior of others. These discoveries were considered too radical until the 1960s, when her results were accepted by her peers. In 1983, McClintock became the first woman to be sole winner of the Nobel Prize in Physiology or Medicine.

MILESTONES

BECOMES BOTANIST
Graduates from Cornell University with an MSc in 1925, then with a PhD in botany in 1927.

JUMPING GENES
Publishes a paper in 1931 proving chromosomal crossover occurs prior to sexual reproduction.

PLANT GENETICS
Accepts a post in Long Island in 1941; makes pivotal discoveries on inheritance in corn.

LEADER IN FIELD
Becomes the third woman to be elected as a member of the National Academy of Sciences in 1944.

GLOBAL RECOGNITION
Receives the National Medal of Science in 1970, after her groundbreaking work is finally accepted.

McClintock observed that genetic information changes as it is passed down generations, resulting in mutations of physical characteristics.

"They thought I was **crazy,** absolutely **mad.**"

Barbara McClintock, 1983

BARBARA
MCCLINTOCK

KONRAD
LORENZ

Austrian zoologist Konrad Lorenz was a pioneer in the field of ethology (the study of animal behavior in natural environments). He is best known for his study of young birds, many of which, he discovered, behave in ways that are both innate (instinctive) and learned from the environment.

Born in Vienna, Austria, Konrad Lorenz was the son of a wealthy surgeon. At his father's wish, he studied medicine, but after qualifying, he went on to gain a PhD in zoology. Animals, especially birds, had always been his passion, and he built up a large menagerie as a child and a young man.

Innate vs. learned behavior

Before Lorenz, biologists made a clear distinction between two types of animal behavior: innate (instinctive) behavior, which is partly genetic, and learned behavior, which is dependent on experience. However, through his study of young birds, Lorenz showed that some behaviors are more complex. He highlighted a type of instinctive behavior that occurs at a particular age and is triggered by the environment.

Birds that can walk as soon as they hatch, such as ducks and geese, form bonds with the first moving object they see within a critical period. Usually this is their mother, but Lorenz discovered that newly hatched greylag geese could be tricked into bonding with him—and even a model train—in a process he later called "imprinting." He observed that the courtship behavior of birds is similarly stage-linked and instinctive, following what he called "fixed-action patterns." Since then, scientists have looked for similar patterns in the behavior of other animals, including humans.

Lorenz conducted his most famous behavioral experiments with greylag geese. The goslings that saw him before any other moving object when they hatched would instinctively bond with him as their "mother," thereafter following him everywhere.

MILESTONES

IMPRINTING TESTS
Carries out his first experiment on geese in 1935, with more to follow on geese and jackdaws.

WARTIME DOCTOR
Serves as a doctor in the German army, is captured by the Soviets in 1944, and is imprisoned for 4 years.

POPULAR SCIENCE
Publishes *King Solomon's Ring* in 1949. Written for a popular audience, it is one of his most famous books.

NOBEL AWARD
Wins the 1973 Nobel Prize in Physiology or Medicine, jointly with Karl von Frisch and Nikolaas Tinbergen.

Theoretical physicist J. Robert Oppenheimer's widespread intellect encompassed molecular physics, quantum theory, and black holes, as well as the mystical reaches of Hinduism. He is most renowned as the mastermind of the Manhattan Project, the US-led atomic weapons program.

Born into a Jewish family in New York City in 1904, J. Robert Oppenheimer was fascinated by minerals, physics, and chemistry as a child. Aged 12, he was invited to lecture at New York's Mineralogical Club, which had assumed the author of the high-level letters it received must be an adult. Oppenheimer graduated at the top of his high school class and enrolled at Harvard in 1922. There he studied Latin, Greek, Eastern philosophy, poetry, and the sciences and majored in chemistry in 1925. But theoretical physics was his real passion, and he went to Cambridge University to study the subject as a postgraduate.

Europe's leading physicists were then developing quantum theory, and Oppenheimer was invited to the University of Göttingen, Germany, to work with eminent physicist Max Born. Together, in 1927, they produced the Born–Oppenheimer approximation, which explained, as Oppenheimer put it, "why molecules are molecules." Extending quantum mechanics beyond single atoms to describe the energy of chemical compounds,

"Science is not everything, **but science is very beautiful."**

J. Robert Oppenheimer, 1966

The Trinity test site in New Mexico was the scene of the world's first nuclear explosion. Oppenheimer (center left) likened the terrible force of the blast to words from the Sanskrit scripture the Bhagavad Gita: "I am become Death, the destroyer of worlds."

J.ROBERT **OPPENHEIMER**

1904–1967

"The physicists have known sin ... a knowledge which they cannot lose."

J. Robert Oppenheimer, 1947

their concept would later prove vital in simplifying equations used to calculate the energy of a molecule. Oppenheimer held postgraduate roles at the California Institute of Technology (Caltech) and at Harvard, as well as in the Netherlands and Switzerland, before settling in the US in 1929 to teach at Caltech and Berkeley.

Nuclear dawn

Oppenheimer's research ranged from energy processes in subatomic particles—including cosmic rays, electrons, and positrons—to black holes and neutron stars. Yet it was the 1939 discovery of nuclear fission by Lise Meitner (see pp.194–197) and Otto Hahn that would definitively shape his career.

When the US entered World War II in 1941, Oppenheimer took up atomic weapons research, which he judged "a good, honest, practical way" to use science. By July 1942, he had emerged as the natural leader of a group of scientists convened by the Manhattan Project, the Allied nuclear-weapons program.

Atomic pioneer

Oppenheimer's ambition and intellect were well suited to the vast scale of the project's interdisciplinary collaboration, and in 1943, he was made director of

VANNEVAR **BUSH**

Successful inventor and engineer Vannevar Bush was an influential figure in US science administration—in particular, the Manhattan Project—during and after World War II.

Bush (1890–1974) was tasked with leading the National Defense Research Committee by President Roosevelt in 1940. He informed the president of the 1941 UK MAUD Report, which had concluded that an atomic bomb was feasible within the expected timescale of World War II, leading directly to the creation of the Manhattan Project. In 1945, he outlined an idea for a desklike computing device, called a memex, that helped inspire Tim Berners-Lee's World Wide Web some 40 years later.

Oppenheimer directed the huge-scale, complex Project Y, which grew to have more than 6,000 staff, including the highest concentration of Nobel laureates ever gathered in one place.

PROJECT Y

Project Y, the secret facility in New Mexico tasked with making the atomic bomb. His leadership and grasp of scientific, military, and engineering detail were integral to the project's success.

Two types of weapons were perfected: one using the "gun-type" uranium method and the other the "implosion-type" plutonium method of triggering a nuclear chain reaction. The former was used in the bomb that obliterated Hiroshima, Japan, on August 6, 1945, and the latter was used in the Trinity test of July 16 and the bomb dropped on Nagasaki, Japan, on August 9, which precipitated Japan's surrender.

Peacetime remorse

After initial euphoria, Oppenheimer was horrified at the devastation the bombs caused. He resigned from the Manhattan Project later in 1945, and instead mixed academia and atomic consultancy. After he opposed the development of the H-bomb, opinion turned against him; he lost his security clearance in 1953 and was dismissed from the US Atomic Energy Commission.

LED A TEAM OF
3,000
ATOM-BOMB
SCIENTISTS

NOMINATED FOR
3 NOBEL
PRIZES BUT
NEVER
WON

Spanish American biochemist Severo Ochoa was awarded the Nobel Prize for discovering a bacterial enzyme that synthesizes ribonucleic acid (RNA)—a molecule that is essential to all forms of life. His findings were partly flawed, but they were valuable, as they revealed more of the process by which genetic information is encoded in cells.

Born in Madrid, Severo Ochoa held posts at European and US universities before becoming professor of biochemistry at New York University, which was renowned for its work in enzymology. In 1954, the year after Francis Crick and James Watson demonstrated the helix structure of DNA (see pp.264–269), Ochoa became interested in deciphering the genetic code and decided to investigate the mechanism for synthesizing nucleic acids.

In 1955, after a complex process of elimination in his laboratory, Ochoa chose the *Azotobacter vinelandii* bacterium to isolate an enzyme that he called polynucleotide phosphorylase. In test-tube conditions, the enzyme was found to synthesize ribonucleic acid, or RNA—a molecule that, like DNA, plays an important role in the transfer of genetic information.

Discovery debated
While Ochoa was conducting his experiments, the US biochemist Arthur Kornberg achieved a similar result with a different bacterium. The enzyme that Kornberg isolated appeared to explain the mechanism for synthesizing DNA. The two men shared the 1959 Nobel Prize in Physiology or Medicine for their discoveries, but soon afterward, observations revealed that the roles assigned to the two enzymes could not be confirmed. Nevertheless, Ochoa's enzyme was helpful in his later work on the genetic code.

Ochoa and his team were delighted to hear of the Nobel Prize nomination for their work, which showed that RNA can be synthesized artificially.

"Polynucleotide phosphorylase may be considered ... the Rosetta Stone of the genetic code."

Severo Ochoa, 1980

SEVERO OCHOA

1905–1993

MARIA

GOEPPERT MAYER

German-born Maria Goeppert Mayer was a theoretical physicist who, in 1963, became the second woman to win the Nobel Prize in Physics. She created the nuclear shell model, which explained how some isotopes (variants of an element) would be more stable than others and why some were likely to be radioactive.

Maria Goeppert Mayer was born in Kattowitz (now in Poland) and studied mathematics and then physics at the University of Göttingen. The university was a leading center of the new field of quantum mechanics, and Goeppert Mayer studied under renowned physicist Max Born. In her doctoral thesis in 1930, Goeppert Mayer predicted that some atoms could absorb two photons (the smallest unit of light) simultaneously, and as a result move to a higher energy state. Her theory was not proven until 1961, when the invention of the laser made it possible to demonstrate the effect in a crystal of the metal europium. Multiphoton absorption has since been used for body scanning and creating polymers (chemicals made of repeating units).

Goeppert Mayer moved to the US in 1930, when her husband accepted a position as assistant professor of chemistry at Johns Hopkins University, and she spent many years as an unpaid researcher at Johns Hopkins and Columbia Universities. During this time, she published a number of scientific papers, including one, in 1935, on a type of radioactive decay called double-beta decay. In 1942, she was invited

MILESTONES

CHANGE OF DIRECTION
Changes course to study physics after attending an inspirational quantum mechanics seminar in 1924.

PAID POST
After years of unpaid work, is offered a senior role in 1946 at Argonne National Laboratory, Illinois.

CATALYST QUESTION
A colleague's question prompts her nuclear shell model; publishes her theory in *Physical Review* in 1950.

NOBEL PRIZE
In 1963, is jointly awarded the Nobel Prize in Physics for her discoveries on nuclear shell structure.

During the Manhattan Project, Goeppert Mayer joined Edward Teller's team at the Los Alamos Laboratory, New Mexico, to assist with the development of the first hydrogen bomb: a powerful bomb that used nuclear fusion and fission.

"Physics is **puzzle solving** ... but of puzzles **created by nature**, not by the mind of man."

Maria Goeppert Mayer

DISCOVERED THAT 2, 8, 20, 28, 50, 82, OR 126 PROTONS AND NEUTRONS MAKE THE NUCLEUS STABLE

FINALLY APPOINTED PROFESSOR OF PHYSICS AT THE AGE OF 54

1 OF JUST 3 FEMALE NOBEL PRIZE-WINNING PHYSICISTS

to join the Manhattan Project, which was building the world's first atomic weapon. She joined a group of scientists at Columbia University who were seeking to separate the isotope U-235 from natural uranium in their research toward building an atomic bomb.

Nuclear shell model

In the late 1940s, Goeppert Mayer was working at the Argonne National Laboratory and the Institute for Nuclear Studies in Chicago. She was studying why some isotopes (different forms of the same element) with certain numbers of protons and neutrons in their nucleus were more stable than others and were therefore more likely to exist. She had found seven such numbers. When the amount of protons or neutrons in the nucleus was one of these seven numbers, the isotope would have a particularly stable nuclear structure, but she could not fully explain why. In 1949, the nuclear physicist Enrico Fermi (see p.196) asked her: "Is there any indication of spin-orbit coupling?" Goeppert Mayer said this question gave her the answer she was

looking for, and the next day she put forward her theory of the nuclear shell model. Electrons were known to exhibit spin-orbit coupling—they orbited the nucleus in pairs, with one spinning clockwise and the other one counterclockwise. Goeppert Mayer reasoned that pairs of protons and neutrons were behaving in the same way in concentric shell-like structures within the nucleus. She described these pairs as being similar to circles of dancers waltzing in different directions and spinning at the same time.

In her theory, Goeppert Mayer stated that when the outer shell of the nucleus was full (containing one of her seven numbers of protons or neutrons), it would be very stable. When it was not full, the nucleus would be less stable. This helps explain why elements such as helium, oxygen, and calcium are common: they have stable nuclei that fit this number pattern. Goeppert Mayer's mathematical background and

EUGENE WIGNER

US professor of mathematical physics Eugene Wigner furthered the theory of the atomic nucleus and elementary particles, for which he shared the Nobel Prize in Physics in 1963.

Wigner (1902–1995) worked at the University of Göttingen. In 1933, he worked out that the binding force of nucleons (protons and neutrons) depends on how close they are together. He is thought to have coined the term "magic numbers" for the stable numbers of nucleons described by Goeppert Mayer. In 1939, Wigner and others encouraged Einstein to send a letter to President Roosevelt, which led to the Manhattan Project.

understanding of quantum physics had enabled her to recognize that spin-orbit coupling was the key to the correct solution: a breakthrough that was hugely important for nuclear physics.

In 1963, Goeppert Mayer was awarded the Nobel Prize in Physics in 1963 with Hans Jensen, who had come up with the same nuclear shell model independently, and Eugene Wigner (see box).

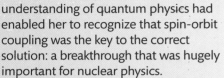

"Winning the [Nobel] Prize wasn't half as exciting **as doing the work itself.**"

Maria Goeppert Mayer

__Goeppert Mayer noticed__ that isotopes with a certain number of protons or neutrons were particularly stable and more abundant. Inexplicable at the time, these numbers became known as "magic." Studying why this was the case led Goeppert Mayer to propose the nuclear shell model.

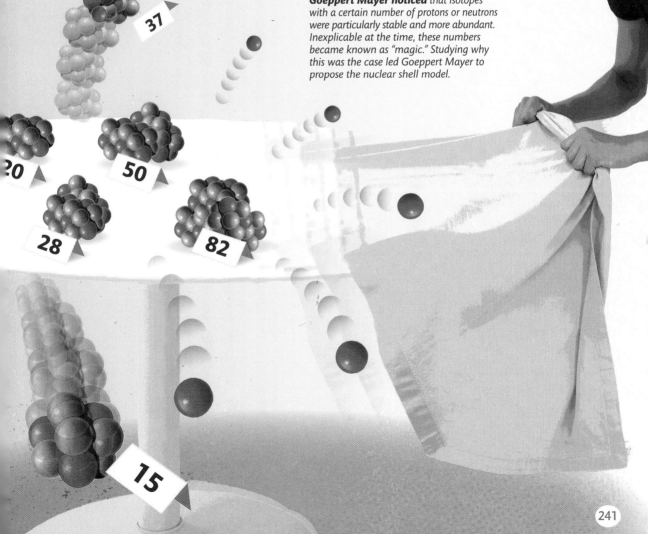

US mathematician, computer scientist, and naval officer Grace Hopper was a pioneer in the field of computer technology. She helped to develop the earliest computer languages, which enabled computers to be programmed with ordinary words.

After graduating from Vassar College, New York, Grace Hopper attended Yale University and gained her PhD in mathematics in 1934. She joined the US Naval Reserves in 1943. During her long career, Hopper would attain the position of rear admiral in the navy and become one of the foremost computer programmers of her generation.

Pioneering technology

In 1944, Hopper was posted to Harvard University, where she helped develop the Mark series of computers. This began with Mark I, which was one of the first electromechanical computers and a prototype of the electronic computer. In 1949, she joined the Eckert-Mauchly Computer Corporation to help build UNIVAC 1, the earliest computer designed for business and administrative use in the US. Soon after, she improved the computer compiler—a program that translated symbolic mathematical instructions into codes that a computer could read directly. This was a key step in the development of computer languages.

Hopper championed the idea of making computers accessible to a nontechnical market, and in 1955, she designed Flow-Matic, the first programming language that responded to English word commands. She then helped develop, refine, and promote COBOL, a business-orientated language that became widely used. With her visionary skills and outstanding technical ability, Hopper simplified programming and enabled broad commercial usage of computer technology.

Unlike her contemporaries, Grace Hopper understood the potential commercial applications of computers and made such usage possible through her ongoing work on compilers.

1906–1992

GRACE
HOPPER

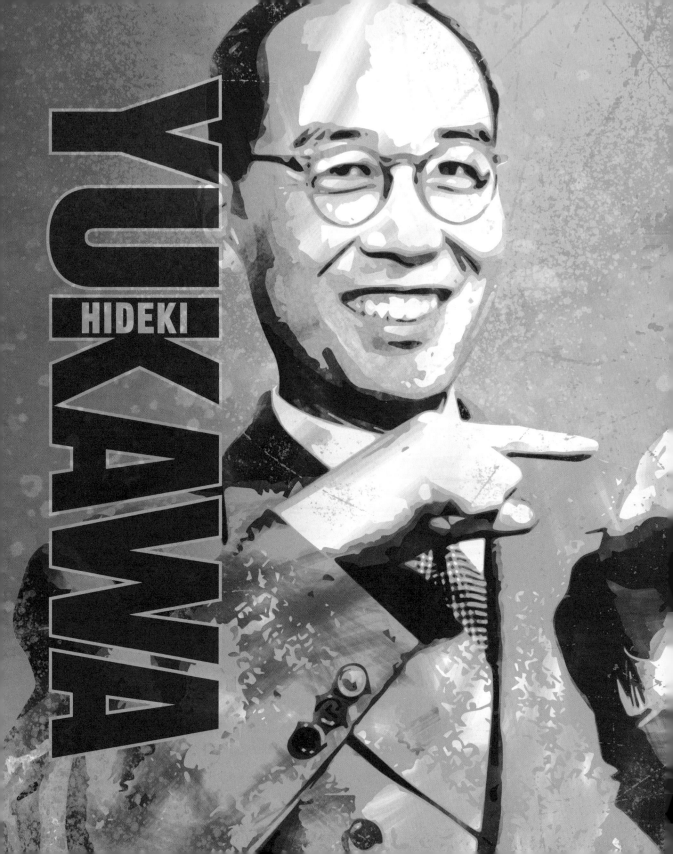

YUKAWA

HIDEKI

Japanese physicist Hideki Yukawa's Nobel Prize-winning work advanced our knowledge of two of the four basic forces in nature—the strong and weak nuclear forces. He proposed the existence of a new kind of subatomic particle, the meson, to explain the actions of protons and neutrons inside atoms.

In 1935, while a physics lecturer at Osaka Imperial University, Hideki Yukawa published a pioneering paper on nuclear forces. The proton had been discovered in 1919, and the neutron in 1932, and physicists knew that atomic nuclei contain both particles. However, it was not clear how the positively charged protons were able to interact within the nucleus without repelling one another. Yukawa's theory predicted that the protons and neutrons interacted by exchanging an intermediary "carrier" particle, which was later called a meson.

Meson's existence confirmed

In 1937, a team of US scientists discovered a new particle in cosmic radiation which had a mass close to that of Yukawa's predicted meson. This appeared to confirm Yukawa's theory, and his reputation spread beyond Japan. However, it was later discovered that a different particle, now called the pion (or pi-meson), was involved in binding the nucleus instead. With Yukawa's work on nuclear forces confirmed, he was awarded the Nobel Prize in Physics in 1949.

MILESTONES

PREDICTS PARTICLE
Claims in his 1935 paper that mesons carry special properties that hold the atomic nucleus together.

AWARDED NOBEL PRIZE
Becomes Japan's first Nobel Laureate in 1949 when awarded the Physics Prize for his meson discovery.

BECOMES DIRECTOR
Appointed first chairman of Japan's Research Institute of Fundamental Physics in 1953.

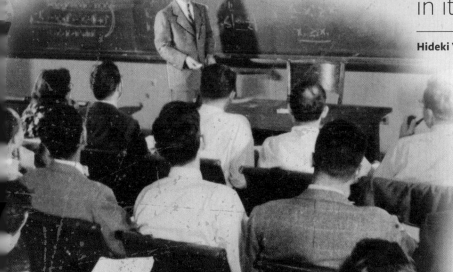

"Nature is simple in its essence."

Hideki Yukawa, 1959

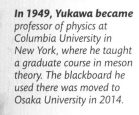

In 1949, Yukawa became professor of physics at Columbia University in New York, where he taught a graduate course in meson theory. The blackboard he used there was moved to Osaka University in 2014.

Two-time Nobel Prize-winning theoretical physicist John Bardeen's knowledge of semiconductors was vital to creating the first working transistor. His expertise made the technology usable, launching a revolution in electronics: transistors are integral to every modern electronic device, from radios to space shuttles.

MILESTONES

SEEKS SOLUTION
Realizes the flaw in the original amplifier design; begins work at Bell Labs to find an alternative in 1945.

HAS BREAKTHROUGH
Finishes his "magic month" of experiments in 1947 and produces the point-contact transistor.

TEAM SPLIT
Clashes with Shockley, who stops him from working on interesting projects, and leaves Bell Labs in 1951.

RADIO SHOCK
Hears he has won the Nobel Prize in Physics in 1956 while listening to the radio.

BCS THEORY
Publishes theory of superconductivity with Leon Cooper and John Robert Schrieffer in 1957.

William Shockley (center) hired John Bardeen (right) and Walter Brattain (left) to work on his team at Bell Labs. Bardeen split from the group following patent disputes with Shockley.

John Bardeen grew up in Madison, Wisconsin, and studied electrical engineering. In 1933, at Princeton University, his tutor Eugene Wigner (see p.240) posed the question "How do electrons inside metals interact?" Bardeen made this the subject of his life's work. For his PhD, he calculated a metal's work function (the amount of energy needed to remove an electron). He started to focus more on superconductivity in 1939. It had been known since 1911 that some metals at temperatures approaching absolute zero became superconductors and allowed electricity to pass through them with no resistance. This implied that their electrons were behaving unusually—but Bardeen was not to discover why until the 1950s.

Semiconductors
In 1945, Bardeen joined the solid state group at Bell Telephone Laboratories. The company used glass vacuum tubes to amplify phone signals, but they were bulky and burned out regularly. The group's task was to invent a solid-state amplifier using a semiconductor: a material that acted as an electrical conductor in some conditions but not in others, making it ideal for amplifying an electric current or

JOHN **BARDEEN**

1908–1991

> "Science is a field which **grows continuously** with **ever-expanding frontiers.** Further, it is truly international in scope."

John Bardeen, 1972

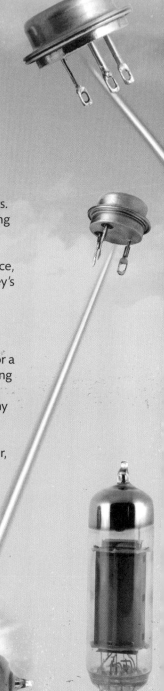

turning it on or off. William Shockley, the scientist in charge, had designed a "field-effect" amplifier: he thought that creating a strong electrical field close to a cylinder coated in the semiconductor silicon would affect its conductivity. When the device did not work, Shockley asked Bardeen and another scientist on the team, Walter Brattain, to investigate.

In an intensive series of experiments, Bardeen, who handled the theoretical aspects of the research, and Brattain, who ran the laboratory, reworked the apparatus. They created a "point-contact" transistor, which used a contact made from thin gold foil to send a small current across the surface of a crystal of germanium (a semiconductor). This changed the semiconductor so that a larger current flowed out through a second gold contact, a hair's breadth away from the first. They had, in 1947, created the first solid-state amplifier (without a vacuum tube). They called it a transistor (a contraction of "transfer and resistor") because of its dual properties of electrical conductance and resistance.

Shockley was angry that he had not been part of the original invention and, working alone, developed an improved design; his much more commercially successful "junction" transistor used a "sandwich" of different semiconductors. Initially adopted as amplifiers in hearing aids and radios, transistors soon found their way into computers, telephones, and almost every other electronic device, making Bardeen, Brattain, and Shockley's work the basis for the 20th century's electronics and computer revolutions.

Superconductors
Bardeen left Bell Laboratories in 1951 for a post as professor of electrical engineering at the University of Illinois. He returned to researching superconductors and why their conductivity increases at very low temperatures. He and two colleagues, Leon Cooper and John Robert Schrieffer, worked out what became known as the BCS theory (from their initials) of

superconductivity in 1957. Bardeen suggested that whereas electrons usually repel each other, at very low temperatures they attract each other. Cooper said that the electrons worked together in what became known as Cooper pairs, and Schrieffer stated that the Cooper pairs moved in a coordinated way, as a single wave, so that the electrical resistance disappeared. In 1972, the three physicists were jointly awarded the Nobel Prize for their theory. Superconductors now have a wide range of applications, including use in powerful electromagnets in magnetic resonance (MRI) machines and maglev trains, which levitate above the tracks.

WILLIAM **SHOCKLEY**

American physicist and inventor William Shockley helped make the first transistor.

Shockley (1910–1989) had a background in radar research and solid state electronics, and was one of the first to work with silicon as a semiconductor. While at Bell Telephone Laboratories, he helped invent the transistor and then developed it further to produce the junction transistor. In 1955, he founded Shockley Semiconductor, one of the first electronics companies in what is now Silicon Valley.

BARDEEN AND BRATTAIN'S **TRANSISTOR** WAS **0.5 IN** (**1.25 CM**) HIGH

PROVIDED THE **FIRST** SUCCESSFUL MODEL OF **SUPER** CONDUCTIVITY

1 OF JUST **4** PEOPLE TO HAVE **WON** THE **NOBEL** PRIZE **TWICE**

Bardeen's transistor started a revolution in the electronics industry. It was favored for its size and reliability over vacuum tubes; most electronic devices would become transistor-based within 25 years.

DOROTHY HODGKIN

A leader in crystallography, British chemist Dorothy Hodgkin revealed the structure of complex molecules, including insulin.

As a child, Dorothy Hodgkin developed a passion for crystals and would analyze garden stones with a mineral-testing kit. She went on to study physics and chemistry at Oxford University, where she learned about using X-ray crystallography to model the structure of a crystal. She obtained a PhD and became an expert in this new field. In 1942, Hodgkin was asked to analyze the crystals of penicillin salts, and after 3 years of research, she succeeded. At the time, penicillin was the largest molecule to have been described using this method. Hodgkin went on to identify the structure of vitamin B12 and, in 1969, the hormone insulin—a structure she had begun studying 35 years earlier. She considered this her greatest discovery. A lifelong champion of social equality, Hodgkin devoted her later years to the advocacy of disarmament and peace.

A pioneer in her field, Hodgkin used sophisticated computer models and X-ray crystallography to determine the complex structure of B12.

The brilliant Indian American astrophysicist Subrahmanyan Chandrasekhar worked out how stars evolve and discovered that white dwarf stars over a certain mass collapse into denser objects.

Born in Lahore, now Pakistan, Subrahmanyan Chandrasekhar was a child prodigy and began a physics degree in Madras (now Chennai), India, at the age of 14. In 1930, he won a scholarship to study for a PhD in astrophysics at Cambridge University. On the boat to England, he determined the maximum mass of a white dwarf, a calculation that would later win him the Nobel Prize.

When a star has used up its nuclear fuel, it first sheds most of its mass, then it starts to collapse under its own gravity. Even stars the size of our Sun will shrink to become "white dwarfs." In 1930, physicists believed that every star is prevented from further collapse by its "degenerate electron pressure"—the outward force caused by its electrons being so densely packed. However, Chandrasekhar realized that stars above a certain size will collapse further, and that this occurs when the star's mass (after its collapse) is 1.44 times greater than our Sun. This figure is called the Chandrasekhar limit.

His theory was ridiculed by his contemporaries, who thought that it was not possible for matter to be compressed to the densities Chandrasekhar predicted. But in 1971, the first black hole was discovered, and Chandrasekhar was vindicated.

MILESTONES

KEY DISCOVERY
Defines the Chandrasekhar limit in "The Maximum Mass of Ideal White Dwarfs," published in 1931.

DEVELOPS JOURNAL
Editor of *The Astrophysical Journal* from 1952 to 1971; he builds it into a leading international publication.

LATE RECOGNITION
Wins Nobel Prize in Physics in 1983, jointly with William Fowler, and finally gains credit for his earlier work.

"In science, **beauty** is often **the source of delight.**"

Subrahmanyan Chandrasekhar, 1975

SUBRAHMANYAN
CHANDRASEKHAR

MILESTONES

EARLY CAREER
Joins the University of California in 1936; makes key contributions to work with subatomic particles.

INTERNATIONAL ACCLAIM
Wins the Nobel Prize in Physics in 1968 for his discoveries in elementary particle physics.

EXTINCTION THEORY
Works with his son, Walter, in 1980 to propose a theory explaining the extinction of dinosaurs.

Luis Alvarez was a US experimental physicist who received a Nobel Prize for his work with subatomic particles. He was also the first to hypothesize that a meteorite caused the extinction of the dinosaurs.

After obtaining a PhD in physics from Chicago University, Luis Alvarez joined the Radiation Laboratory of the University of California, Berkeley, as an experimental physicist. He specialized in subatomic particles and, in 1937, revealed the process by which a radioactive atom transforms into a new element (K-electron capture). This occurs when the atom's nucleus captures an orbiting electron, which changes the atomic structure and creates a new element. Using a particle accelerator, he also proved that the helium-3 isotope is stable (not radioactive). During World War II, Alvarez developed radar systems and a detonation device for plutonium bombs.

In the late 1950s, he designed a hydrogen bubble chamber, which made it possible to observe and analyze subatomic particles outside the accelerator. This allowed Alvarez and others to discover many new particles and earned him the Nobel Prize in Physics. In 1980, Alvarez theorized that a meteorite led to the extinction of the dinosaurs, after finding iridium (a metal found in meteorites) in rock strata 65 million years old. Support for his theory came a decade later.

Alvarez's dinosaur extinction theory was based on analysis of rock strata from Italy. With the discovery of a vast crater in Mexico in 1990, his theory was confirmed.

LUIS ALVAREZ

Chinese American nuclear physicist Chien-Shiung Wu, a female trailblazer in the scientific world, gained international acclaim for disproving the universal nature of a fundamental law of physics.

Chien-Shiung Wu moved from China to the US to study in 1936. She earned her PhD in nuclear physics from the University of California, Berkeley, where her work included splitting uranium atoms to create radioactive isotopes. Wu then began teaching at Princeton as the institute's first female lecturer before moving in 1944 to work on the Manhattan Project (working toward the atomic bomb) at the University of Columbia, where she helped develop radiation detectors and Geiger counters. In 1956, she carried out the Wu Experiment, which resulted in her major contribution to science. It demonstrated that a fundamental principle of physics (conservation of parity) is not universally valid. This principle was believed to govern the behavior of all subatomic particles until Wu's experiment proved it did not apply in all situations. She won global acclaim for her discovery and helped secure a Nobel Prize for fellow physicists Yang Chen-Ning (see p.284) and Tsung-Dao Lee, whose theory she confirmed.

"This **wonder** ... can be the reward of a lifetime."

Chien-Shiung Wu, 1957

CHIEN-SHIUNG WU

A pioneer in computing and artificial intelligence, mathematician and logician Alan Turing developed code-breaking machines that helped to turn the tide of World War II. He wrote the blueprint for how computing machines could be programmed to perform any task, heralding the advent of the computer age.

MILESTONES

THE TURING MACHINE
Outlines a "universal computing machine" in his 1936 paper "On Computable Numbers."

CRACKING CODES
Plays an instrumental role from 1938 in mathematical and computational methods of code-breaking.

EARLY PROGRAMMING
Compiles his Abbreviated Code Instructions in 1947, the first-ever computer programming language.

THE TURING TEST
Writes a test to assess artificial intelligence in "Computing Machinery and Intelligence" in 1950.

COMPUTING FIRST
Programs the first commercially marketed digital electronic computer in 1951.

Born in London in 1912, Turing was described as a genius by his headmistress at the age of 9, and demonstrated an early interest in mathematics and complex chess problems. He read mathematics at King's College, Cambridge, gaining a undergraduate degree in 1934 and being elected Fellow in 1935 for a thesis on probability theory. In 1936, he turned to the "decision problem," an unresolved question of logic posed by David Hilbert. Turing succeeded in solving the problem, although he was narrowly beaten to it by US logician Alonzo Church. However, the important thing was Turing's method: he solved the problem using a thought experiment about a machine that could be programmed to perform any task—a computer.

The universal computing machine
Turing imagined a machine that could read and write numbers and symbols printed on a long strip of paper, then perform a predetermined task according to an algorithm—a set of rules.

"Machines take me by surprise with great frequency."

Alan Turing, 1950

At Bletchley Park, Turing was the scientific genius behind the vital work of decrypting German codes. The extent of his role was kept secret until the 1970s, and only revealed in full 50 years after his death.

ALAN TURING

Turing said of his machine: "We have only to regard the rules as being capable of being taken out and exchanged for others and we have something very akin to a universal computing machine." Although the technology did not yet exist to build one, the "Turing machine" hinted at the limitless potential of the future field of computer science.

Code-breaker

After 2 years in the US at Princeton University—where he gained his PhD—Turing was approached by the British government's Code and Cypher School in 1938. The school had been set up to decipher the German military's coded radio messages—the trouble was, the Germans were using an encryption device known as Enigma, which scrambled their messages in different ways each day. Indeed, the Germans even boasted that the Enigma code was unbreakable. With this daunting task ahead of him, Turing left Cambridge—where he had begun building a mechanical computing device—and joined the Cypher School at Bletchley Park. He arrived on September

4, 1939, the day after Britain had declared war on Germany. He worked in a small team of mathematicians and code-breakers who had already enjoyed some success in deciphering Enigma messages.

Using logic and probability, Turing realized that to find out all the possible combinations of the Enigma code, he had to build a computer that was capable of replicating 60 Enigma machines. To that end, he and colleague Gordon Welchman increased the power of the "Bombe"—a computer constructed by Polish cryptographers in 1938. The first improved Bombe was ready in August 1940, and shortly after that, the Enigma messages started to be deciphered. By the end of the war, more than 200 Bombes were operating day and night at Bletchley Park, even cracking the more complex U-boat ciphers.

Digital computing

Turing was also on hand at Bletchley to observe the development in 1944 of the "Colossus" code-breaking machines, the first programmable digital electronic computers. He immediately saw the

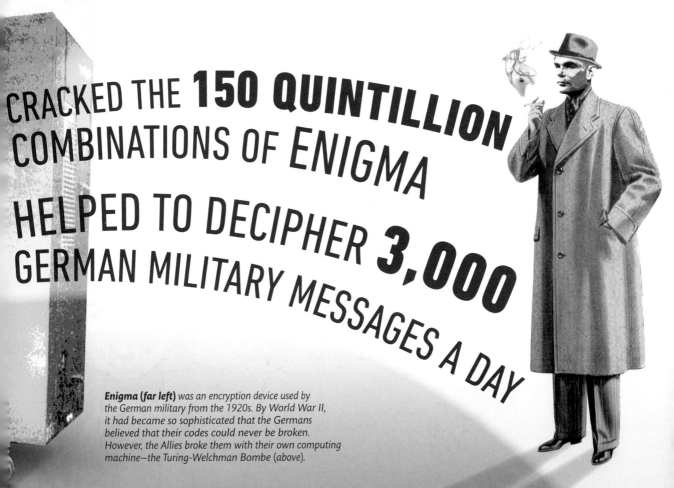

CRACKED THE **150 QUINTILLION** COMBINATIONS OF ENIGMA

HELPED TO DECIPHER **3,000** GERMAN MILITARY MESSAGES A DAY

Enigma (far left) was an encryption device used by the German military from the 1920s. By World War II, it had became so sophisticated that the Germans believed that their codes could never be broken. However, the Allies broke them with their own computing machine—the Turing-Welchman Bombe (above).

Born in Berlin, German engineer Zuse designed and built a programmable computer while working in complete isolation, making the prototype machine in his parents' apartment.

Zuse (1910–1995) conceived the idea of a computer to automate the laborious calculations required in his early career as a civil engineer. In 1936, he left his job at an aviation firm and began constructing his computer, creating the Z1 in 1938. Zuse worked for the German military during World War II, completing the Z3—the world's first fully programmable working computer—in 1941. He went on to develop one of the first high-level programming languages, "Plankalkül."

memory, just like a modern computer. Much to Turing's frustration, it was never built, though the 1950 Pilot ACE was a smaller version of his design.

Machine intelligence

After moving to Manchester University in 1948, Turing led the programming team for the Ferranti Mark I—the world's first commercially available digital electronic computer. While there, he explored how machines could be said to think, using a thought experiment in which a computer tries to convince an interviewer that it is human: success meant it was intelligent. The Turing test, as it became known, was influential in Artificial Intelligence, an emerging field pioneered in the US.

Turing also demonstrated a remarkable range of academic interests. He was fascinated by neurology and biology and wrote a 1952 paper on mathematical morphogenesis, the theory of how form and pattern—such as a tiger's stripes—arise in nature.

In the same year, Turing was convicted of being sexually involved with a man, which was illegal at the time. In 1954, at the age of 41, he committed suicide. In 2013, Turing was granted a royal pardon.

potential for the quicker electronic technology to realize his 1936 idea of a universal machine. With the war over, Turing worked for the UK's National Physical Laboratory from 1945, where he submitted detailed plans for an Automatic Computing Engine (ACE) that could be programmed via instructions stored in an electronic memory. This meant that unlike the cumbersome Colossus, which had to be rewired to perform each new task, ACE could switch from numerical work to algebra or chess-playing simply by accessing a different program in its

DEVISED A METHOD FOR ASSESSING ARTIFICIAL INTELLIGENCE

CREDITED WITH TAKING 2 YEARS OFF THE DURATION OF WORLD WAR II

IT IS ESTIMATED THAT CRACKING ENIGMA SAVED MORE THAN 14 MILLION LIVES

"I BELIEVE THAT AT THE END OF THE CENTURY THE USE OF WORDS AND GENERAL EDUCATED OPINION WILL HAVE ALTERED SO MUCH THAT ONE WILL BE ABLE TO SPEAK OF MACHINES THINKING WITHOUT EXPECTING TO BE CONTRADICTED."

Alan Turing
"Computing Machinery and Intelligence," *Mind* journal, 1950

DIRECTORY

During World War II, governments increased funding and support for research, hoping to gain tactical advantages. Intense competition among scientists racing to make new findings often led to valuable discoveries for civilians, as well as the military, and continued after the war was over.

WOLFGANG PAULI
1900–1958

Austrian-born Swiss and American theoretical physicist Wolfgang Pauli is best known for his work on atomic structure. He created the Pauli exclusion principle, which states that no two electrons in an atom can exist in exactly the same state at the same time. Pauli (and, independently, German theoretical physicist Arnold Sommerfeld) devised an atomic model that explained the electrical and thermal properties of metals. Pauli was also the first to propose the existence of the subatomic particle the neutrino. He was awarded the Nobel Prize in Physics in 1945.

THEODOSIUS DOBZHANSKY
1900–1975

Russian population geneticist Theodosius Dobzhansky was a central figure in the field of evolutionary biology. Born in Ukraine, he emigrated to the US in 1927. Largely through experiments with fruit flies, he helped to bridge the gap between genetics and evolution. He showed that mutation, in creating genetic diversity, supplied the raw material for natural selection to act on. In this way, he helped reconcile the findings of genetics with Charles Darwin's theory of evolution, which placed emphasis on natural selection.

HOWARD AIKEN
1900–1973

American electrical engineer Howard Aiken was responsible for building the Harvard Mark I computer in 1944. Developed for use in World War II, this part-electric, part-mechanical computer showed that programmable machines could crack large-scale problems systematically, reliably, and automatically; it helped to set the stage for today's computers. Aiken worked closely with Grace Hopper, one of the originators of computer programming.

WERNER HEISENBERG
1901–1976

Born in Würzburg, southern Germany, in 1901, Werner Heisenberg studied mathematics and physics at the universities of Munich and Göttingen, meeting his future collaborator Niels Bohr in Göttingen in 1922. Heisenberg is best known for his work on the Copenhagen interpretation (which looks at the way quantum systems, governed by the laws of probability, interact with the large-scale world) and the uncertainty principle (which says only the speed or the position of a particle is known, not both). He also contributed to quantum field theory and worked out his own theory of antimatter. At age 32, he was awarded the Nobel Prize in Physics, making him one of its youngest recipients.

LINUS PAULING
1901–1994

Chemist Linus Carl Pauling was born in Portland, Oregon, and was the winner of two Nobel Prizes—one for Chemistry and the other for Peace (awarded for his attempts at mediating between the US and Vietnam). Pauling was an early proponent of quantum mechanics in chemistry, and his book *The Nature of the Chemical Bond*, published in 1939, is one of the most influential chemistry texts ever published. He set out five rules for finding the structure of ionic crystals and described how atomic orbitals in covalent compounds combine to form molecular orbitals. According to Pauling, when two atoms form a covalent bond, they share electrons but their position is not fixed; instead, the electrons can move to a more favorable position to bond.

KATHLEEN LONSDALE
1903–1971

Working under the direction of William Henry Bragg at the Royal Institution, London, Kathleen Lonsdale developed

X-ray techniques to study the structure of chemical compounds. In 1929, she worked out that the carbon atoms in benzene were arranged in a regular hexagonal ring. She went on to apply X-ray crystallography to the study of bladder stones and drugs used to relax muscles. In 1945, she became one of two women admitted as fellows of the Royal Society, London, which had previously only admitted men.

JOHN VON NEUMANN
1903–1957

Brilliant Hungarian-born mathematician John von Neumann worked on the design and development of early high-speed electronic digital computers. From 1930, he was based at Princeton, New Jersey, where—with American meteorologist Jule Charney—he put together the first computer-based weather predictions. In 1945, he worked out a model for a computer that could store programs, as well as data. All computers since then have been based on this model, which is known as von Neumann architecture.

WENDELL STANLEY
1904–1971

Born in Indiana, Wendell Stanley determined the molecular structure of viruses by using X-ray diffraction. He went on to share a Nobel Prize in Chemistry in 1946 for his research on the tobacco mosaic virus. During World War II, his insight into viruses as the cause of infectious disease aided the development of a vaccine against influenza. Shortly after the war, he moved to the University of California, Berkeley, and in 1954, he and his collaborators crystallized the polio virus—a major step in enabling researchers to develop a vaccine against the disease.

RITA LEVI-MONTALCINI
1909–2012

After struggles with her father, who believed women should stay at home, Rita Levi-Montalcini studied medicine at the University of Turin and went on to make important discoveries about nerve growth factors (NGFs). She moved to the US in 1947, and was made a professor at Washington University. Initially largely ignored, NGFs were later understood to be important in several diseases, including Alzheimer's, infertility, and cancer. Along with biochemist Stanley Cohen, who helped isolate an NGF, Levi-Montalcini was awarded the 1986 Nobel Prize in Physiology or Medicine.

RUBY PAYNE-SCOTT
1912–1981

Australian Ruby Payne-Scott had a short but productive career as a radio physicist in the 1940s and '50s. In June 1941, she was one of only two women physicists to be employed by the Radiophysics Laboratory at the University of Sydney. She helped develop radars for Australia's coastline in World War II, and from 1946 to 1951 was part of the team that developed a means of measuring radio emissions from the Sun and stars.

JONAS SALK
1914–1995

American virologist Jonas Salk developed the first safe and effective vaccine for polio. His vaccine used the killed virus—no live virus was injected. Early trials showed this vaccine gave immunity against the disease, but it was not long-lasting. In 1954, large-scale trials began. More than 1 million children were injected with the vaccine, which proved up to 90 percent effective against paralytic polio. At the same time, Albert Sabin was developing an oral vaccine against polio using live, attenuated (weakened) viruses. After another major trial, this vaccine was deemed successful, too. Easier to administer than the Salk vaccine, Sabin's oral vaccine had largely replaced Salk's by the early 1960s.

LYMAN SPITZER JR.
1914–1997

Renowned US astrophysicist Lyman Spitzer Jr. was the first scientist to conceive the idea of space telescopes. He had highlighted the problem of detecting nonvisible radiation through Earth's atmosphere. The solution that he proposed, in 1946, was to put a telescope into space. He understood the obstacles to such a proposal, which included the challenges of space travel and of designing an instrument capable of operating in space by remote control from Earth. In 1977, his campaign for a space telescope paid off, and Congress granted federal funding for the Hubble Space Telescope.

ROBERT EDWARDS
1925–2013

British physiologist Robert Edwards developed the techniques that allowed the first successful "test-tube" pregnancy. Edwards began working on fertilization in 1955 and spent 20 years creating the right conditions for an egg to be fertilized artificially, outside the womb. He began collaborating with surgeon Patrick Steptoe in 1968. Steptoe harvested eggs from female volunteers using keyhole surgery. The first IVF baby was born in July 1978, but Edwards had to wait over 30 years to be awarded a Nobel Prize in Physiology or Medicine in 2010.

7

THEORIES OF EVERYTHING

1950–PRESENT

Nobel Prize-winning British biophysicist Francis Crick is celebrated for his joint 1953 discovery of the structure of DNA, one of the most significant breakthroughs in scientific history.

Francis Crick studied physics at University College London, then began graduate work there on the viscosity of water. During World War II, he was recruited by the Royal Navy to help develop magnetic and acoustic mines, and afterward decided to retrain in biology. In 1949, he joined the Cavendish Laboratory in Cambridge and began a new PhD on the X-ray diffraction of proteins. It was here, in 1951, that Crick met James Watson, with whom he would unlock one of science's greatest mysteries. They had an immediate connection and collaborated in the race to discover the molecular structure of DNA. In 1954, Crick completed his PhD, and went on to make key contributions in the ongoing study of genetic code. He moved to California in 1976, where he studied developmental neurobiology and human consciousness.

MILESTONES

CAMBRIDGE WORK
Starts PhD at the Cavendish Laboratory in 1949; meets Watson in 1951, and starts investigating DNA.

DOUBLE HELIX
Co-publishes a landmark paper in *Nature* on April 25, 1953, revealing the structure of DNA.

PRESTIGIOUS HONORS
Receives awards, including the Nobel Prize in 1962, and medals from the Royal Society in 1972 and 1975.

GAINS PROFESSORSHIP
Becomes Distinguished Research Professor at Salk Institute for Biological Studies, California, in 1977.

"The **genetic code** is ... the **key** to molecular biology."

Francis Crick, 1966

FRANCIS CRICK

US geneticist and biophysicist James Watson launched a new era in biological research with his seminal co-discovery of the double helix structure of DNA, for which he was awarded a Nobel Prize.

James Watson enrolled at the University of Chicago aged 15. He gained a degree in zoology in 1947, then in 1950 completed his PhD in virus research. In 1951, he moved to the Cavendish Laboratory in Cambridge, where he formed a tight and productive partnership with Francis Crick. After publishing their groundbreaking paper on DNA structure, Watson moved to California briefly before returning to the Cavendish to work with Crick on virus construction. He lectured at Harvard, becoming professor of biology in 1960, and published several key biological texts. In 1989, he led the Human Genome Project, an international program that aimed to identify and map all the genes of a human chromosome.

MILESTONES

VIRAL EXPERT
Conducts virus research at Indiana University in 1950, and gains vital knowledge to aid DNA investigation.

DNA BREAKTHROUGH
Builds successful model to demonstrate the molecular structure of DNA with Crick in 1953.

BIOLOGY LECTURER
Is a Senior Research Fellow at Caltech from 1953 to 1955; moves to Harvard in 1956 and teaches biology.

BIOLOGICAL RESEARCH
Directs the leading center of molecular research in 1968; leads the Human Genome Project in 1989.

"In science there is only **one answer** and that has to **be correct.**"

James Watson, 1987

JAMES
WATSON

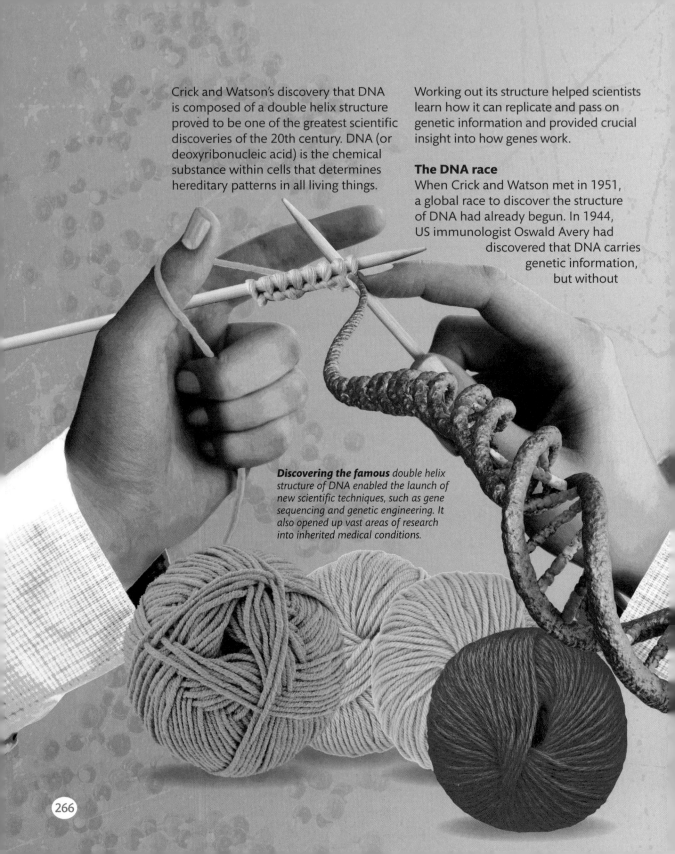

Crick and Watson's discovery that DNA is composed of a double helix structure proved to be one of the greatest scientific discoveries of the 20th century. DNA (or deoxyribonucleic acid) is the chemical substance within cells that determines hereditary patterns in all living things.

Working out its structure helped scientists learn how it can replicate and pass on genetic information and provided crucial insight into how genes work.

The DNA race

When Crick and Watson met in 1951, a global race to discover the structure of DNA had already begun. In 1944, US immunologist Oswald Avery had discovered that DNA carries genetic information, but without

Discovering the famous double helix structure of DNA enabled the launch of new scientific techniques, such as gene sequencing and genetic engineering. It also opened up vast areas of research into inherited medical conditions.

knowing DNA's structure, the study of heredity could not advance. By the 1950s, key players in the race were US chemist Linus Pauling, and Rosalind Franklin and Maurice Wilkins at King's College London.

Scientists knew that DNA could pass on genetic information and make copies of itself. They also knew that the molecule was made up of various components: a "backbone" composed alternately of phosphoric acid and sugar, and four chemical "bases" called adenine (A), thymine (T), guanine (G), and cytosine (C). What was not clear was how these components fit together or how the molecule could replicate.

Dynamic duo

Crick and Watson applied themselves to the investigation with single-minded focus. They compiled evidence from other scientists and pooled their own combined knowledge of disciplines including viral genetics, physics, and X-ray crystallography to interpret the data. In a break from convention, instead of conducting exhaustive experiments, they built large 3D models to try and replicate DNA's structure according to its known chemical properties. After two failed attempts, their success was made possible after obtaining data belonging to Rosalind Franklin (see pp.270–273).

SYDNEY **BRENNER**

South African biologist Sydney Brenner was one of the first to see Watson and Crick's DNA model.

After completing a PhD at Oxford University, Brenner (1927–2019) produced leading research into developmental genetics and molecular biology. In 1961, he discovered the encoding process for the amino acids of a protein with Crick. Brenner is best known for his work on organ development and cell apoptosis (programmed cell death) through his study of transparent nematodes, for which he jointly received a Nobel Prize in 2002.

"We have **discovered** the **secret of life."**

Francis Crick, 1953

MADE THEIR
DNA DISCOVERY
AFTER ONLY
18
MONTHS
OF RESEARCH

BUILT THEIR
**FINAL
DNA**
MODEL IN
3 DAYS

PUBLISHED THEIR
**GROUND-
BREAKING
WORK**
IN A PAPER
JUST **1**
PAGE LONG

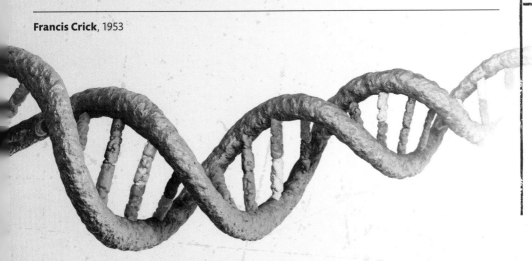

"This structure has features which are of **considerable biological interest.**"

Francis Crick and James Watson, 1953

An expert X-ray crystallographer, Franklin had produced a clear image of DNA—Photograph 51—showing its double helix structure. Franklin was working alongside Maurice Wilkins, but their relationship was hostile, and Wilkins showed Photograph 51 to Crick and Watson without her knowledge. Watson and Crick had already experimented with helical structures but without success. Franklin's image helped them confirm the double helix and complete their model.

Double helix

The double helix structure of DNA resembles a twisted ladder, with the phosphoric acid and sugar strands acting as the ladder's uprights and the paired "bases" forming the rungs. What was crucial was the pairing of the bases—Erwin Chargaff, an Austrian scientist, had previously noted that the A and T bases always appeared in equal ratios, as did the G and C bases. Watson and Crick realized that each rung of the ladder must be composed of two bases (either A and T or C and G) bonded together and that the order of the bases formed the code for genetic information. They deduced that when the ladder splits in half—this is often described as the molecule "unzipping"—each "half ladder" bonds with newly formed DNA components to form two new genetically identical double helixes.

Scientific breakthrough

In March 1953, Crick and Watson completed the double helix model of DNA, and they published their discovery in a landmark paper in April. Franklin's contribution to their work received only passing acknowledgment. Considered one of the greatest scientific discoveries of the 20th century, Crick and Watson's work marked a turning point in our understanding of genetics and paved the way for huge advances in molecular biology. In 1962, along with Maurice Wilkins (Franklin having died in 1958), they were awarded the Nobel Prize in Physiology or Medicine.

Crick (right) and Watson made their famous 3D model out of various materials, including paper and metal plates. ▶

ROSALIND

FRANKLIN

Rosalind Franklin was a British chemist whose revelatory work was vital to the discovery of the structure of DNA. Yet it was largely overlooked in her lifetime.

Born in London in 1920, Rosalind Franklin graduated from the University of Cambridge in 1941 with a degree in physical chemistry. She then took a research post studying the chemical structure of coal and graphite before gaining her doctorate from Cambridge in 1945. Moving to Paris in 1947, Franklin studied X-ray crystallography—using X-rays to examine the crystalline form of a substance and to determine its molecular structure. It was her expertise in this field that enabled her to make a crucial contribution to one of the biggest scientific conundrums of the 20th century.

The DNA race
During the 1950s, a number of scientists across the globe were engaged in a "race" to discover the molecular structure of deoxyribonucleic acid (DNA), the chemical found inside all living things, which encodes genetic information. In Cambridge in 1951, James Watson and Francis Crick were attempting to

In 1951, Franklin began photographing strands of DNA at King's College, London, where a group of physicists, biologists, and biochemists were helping pioneer biophysics (using physics to study biological molecules).

MILESTONES

CHEMICAL RESEARCH
Becomes a researcher for BCURA, the British Coal Utilization Research Association, in 1942.

NEW CAREER
Trains under French engineer Jacques Mering in 1947 and becomes an X-ray crystallographer.

KING'S COLLEGE, LONDON
Asked to head X-ray research in 1951, without the knowledge of post-holder Maurice Wilkins.

DNA BREAKTHROUGH
Produces the first sharp image of crystalline DNA in 1952, which reveals its double helix shape.

PIONEERING WORK
Moves to Birkbeck College, London, in 1953 and leads groundbreaking studies into major crop viruses.

"Science and everyday life cannot and should not be separated."

Rosalind Franklin, 1940

"The **most beautiful X-ray** photographs ... **ever taken.**"

J. D. Bernal, 1958

***Photograph 51 produced** the first clear image of the crystalline structure of DNA, and has been heralded as one of science's most important photographs.*

construct their own 3D model of DNA. Meanwhile, at King's College, London, Franklin was recruited to take over DNA research using X-ray techniques. This caused tension with the former leader of the research team, Maurice Wilkins.

After refining her X-ray techniques, in 1952 Franklin produced the first clear image of DNA, labeled Photograph 51. From the "X" shape that her image revealed, she noted that the structure of DNA was a helix (spiral) and that it was formed of two strands, not three as previously believed. Franklin also proposed the "backbone" of DNA lay on the outside of the molecule.

In early 1953, and without Franklin's knowledge, Maurice Wilkins showed Photograph 51 to Crick and Watson, as well as Franklin's unpublished thoughts about the shape of DNA. These helped the pair confirm DNA's double helix structure—they completed their model and published the results in April 1953, but Franklin's key contributions went unrecognized.

Beyond DNA

Franklin moved to Birkbeck College, London, in 1953 to study the crystalline structure of viruses. Applying her skill as a crystallographer, she proposed a groundbreaking new model for the molecular structure of the tobacco mosaic virus (TMV). This was proven to be correct,

but not until after Franklin's untimely death in 1958. Before she died, she also studied the crystalline structure of other plant viruses and the human virus polio. Her work on polio was recognized when its crystalline structure was confirmed in 1959. In 1962, Watson, Crick, and Wilkins won a Nobel Prize for their work on the structure of DNA. Franklin was ineligible, as the prize was not awarded posthumously. Today, Franklin is credited with having made huge advances in the field of virology (the study of viruses) and for her significant contribution to one of the greatest scientific discoveries of all time.

MAURICE **WILKINS**

Born in New Zealand but raised in the UK, Maurice Wilkins was a Nobel Prize-winning biophysicist best known for his X-ray studies that helped determine the structure of DNA.

Having studied at Cambridge and Birmingham universities, Wilkins (1916–2004) worked on the Manhattan Project (developing the nuclear bomb) during World War II before joining the biophysics unit at King's College, London, in 1946. Here, he used X-ray imaging to study and gather data on DNA fibers, the latter in an unhappy partnership with Rosalind Franklin. In 1962, Wilkins shared a Nobel Prize in Physiology or Medicine with Francis Crick and James Watson for his work on the structure of DNA. However, his contribution is not well known, and Wilkins called his autobiography *The Third Man of the Double Helix*. He became director of the biophysics unit in 1970, and later emeritus professor of King's College.

PHOTOGRAPH 51 TOOK **60 HOURS** TO PRODUCE IN AN **X-RAY**

RECEIVED A **$13,000** GRANT TO STUDY THE **POLIO VIRUS**

BEQUEATHED **$4,000** TO SUPPORT A FELLOW SCIENTIST, **AARON KLUG**, IN HIS WORK

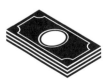

Two-time Nobel laureate Frederick Sanger was an experimental biochemist who sequenced the components of the three fundamental polymers of life: proteins, RNA, and DNA. His technique for sequencing DNA molecules revolutionized molecular biology and heralded a new era in medicine, gene therapy, and genetic manipulation.

Frederick Sanger was born in 1918 in Gloucestershire, UK, the son of a doctor who ran a rural general practice. As a child, he was influenced by his brother's interest in natural history, but equally formative were his father's Quaker beliefs, which gave him a strong sense of truth and conscience—philosophies that would later infuse his scientific career. Sanger went to Cambridge University in 1936 to study natural sciences, and it was there that he became interested in biochemistry; he graduated in 1939 with an undergraduate degree.

Scientific path

Sanger then decided to study advanced biochemistry. His parents had both died while he was an undergraduate; using some of the money he had inherited from them, in 1940, he began a self-funded PhD on the metabolism of amino acids, the basic chemical building blocks of proteins. Exempted from military service during World War II due to his Quaker pacifism, he instead conducted applied research on nitrogen uptake in potatoes for the government. He received his PhD in 1943.

Inspired by the collective momentum of experimentation and laboratory work, Sanger later cited the methodology of research as his greatest strength. He joined a group of scientists in Cambridge who were studying proteins and spent the next decade painstakingly attempting to decipher the complete amino acid sequence of insulin. The sequence was completed in 1955. Sanger also revealed that each protein possesses a unique set of amino acids and distinct 3D structure. Insulin

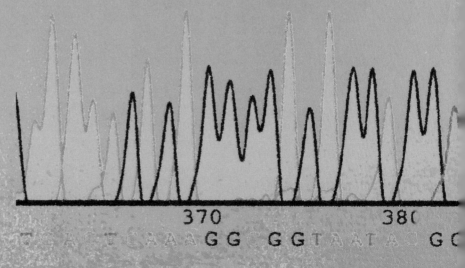

An ingenious experimentalist, Sanger developed novel day-to-day laboratory techniques, such as micromethods for manipulating tiny volumes of chemicals.

FREDERICK SANGER

1918–2013

FIRST PERSON TO WIN **TWO** NOBEL PRIZES IN CHEMISTRY

SEQUENCED 48,502 BASE PAIRS OF DNA BY 1982

was the first protein to be sequenced, and there followed a surge in protein research, which later enabled the development of synthetic insulin for treating diabetes.

Genetic research

Following the award of the Nobel Prize in Chemistry in 1958 for his work on insulin, Sanger had the freedom to pursue his most ambitious research goals and, in 1962, he moved to the Laboratory of Molecular Biology in Cambridge. Having caught what he called "the sequencing bug," he now turned his attention to exploring the structure of the body's genetic nucleic acids, RNA and DNA.

Sanger's Cambridge colleagues Francis Crick and James Watson had discovered the double-helix structure of DNA—the master chemical containing instructions for cell growth and function, as well as inheritable genetic characteristics—in 1953. Sanger aimed to untangle this structure to reveal the bases in specific strands of DNA, each of which carried distinct genetic code. Beginning with RNA, a smaller and less complex single-stranded nucleic acid that relays instructions and creates proteins

> **"I had 20 years when I could just do what I wanted."**
>
> **Frederick Sanger**, 2001

according to the information encoded in DNA, Sanger developed a sequencing method using radioactive isotopes to "label" individual fragments and build up the sequence by looking at the areas of overlap between them. By 1967, he had a complete sequence of RNA from *E. coli* bacteria.

Unlocking the code

In the early 1970s, Sanger moved on to DNA, which was harder to sequence due to its large size and double-helix structure. The sequencing breakthrough came in 1977, when his team pioneered a method of isolating DNA fragments via dideoxy chain-termination—using a molecular inhibitor to prevent DNA

strands from extending. These strands could then be ordered from the shortest to the longest so the base sequence could be read. This rapid, accurate way of sequencing enabled Sanger to map the first human genome, the DNA of cell mitochondria, in 1981, and led the way for sequencing all 3 billion base pairs of the entire human genome.

Sanger's sequencing method
enabled fragments of DNA to be isolated, cloned, and reattached to create modified genes to treat—or be resistant to—genetic disorders.

JOHN **SULSTON**

English biologist John Sulston was the first director of the Wellcome Sanger Institute genome research center and shared the 2002 Nobel Prize for work on programmed cell death.

The most important research carried out by Sulston (1942–2018) centered on *Caenorhabditis elegans*, a 0.03-in (1-mm) nematode worm. This worm has an exact number of cells, so it is ideal for studying cell differentiation, division, and in some cases programmed death. Sulston pushed for the genetic sequencing of *C. elegans*, which in 1988 became the first animal to have its entire genome sequenced. This led directly to the Human Genome Project (1990–2003), in which Sulston was a key contributor.

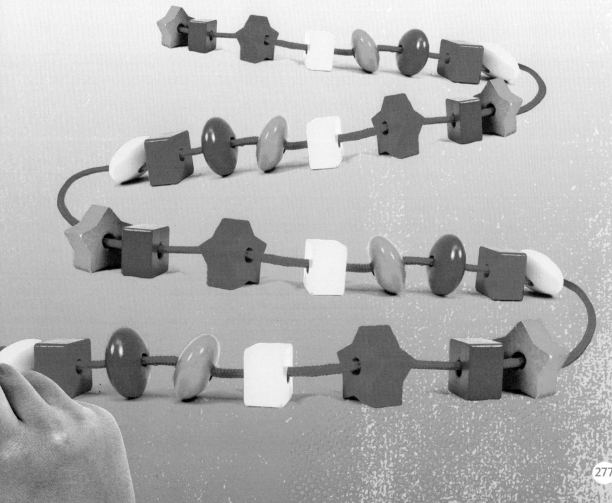

RICHARD **FEYNMAN**

Richard Feynman was an exceptional physicist and inspirational teacher who made the most esoteric of subjects—quantum theory—seem engaging, even entertaining. His subatomic ideas marked the beginning of a new era for physics and technology.

Born in New York to parents of Lithuanian Jewish descent, Richard Feynman did not learn to talk until he was 3 years old but had a talent for working with gadgets and showed a prodigious capacity for mathematics: in high school, he experimented with his own mathematical theories. Refused admission to Columbia University because it had already filled its entrance quota for Jews, he instead attended the Massachusetts Institute of Technology (MIT), studying first mathematics, then physics. In 1939, having gained his degree, he sat the graduate entrance exams for Princeton University and achieved a perfect score in physics—something no one had ever done before. He completed his PhD in quantum mechanics in 1942. During this time, he was asked to participate in the Manhattan Project: the Allied nuclear-weapons program. As he was concerned about the possibility of a Nazi victory in the war, he eventually agreed.

In 1945, Feynman was appointed professor of theoretical physics at Cornell University, where he started his most important work, on quantum electrodynamics (QED), an arcane field theory that had first emerged in the 1920s. QED described the interactions of electromagnetically charged particles in terms of an exchange of photons—packets of light, or "quanta"—of electromagnetic energy. However, there were significant mathematical flaws in the early

Feynman conducted secret bomb research for the Manhattan Project at Los Alamos National Laboratory, New Mexico. He and many other scientists involved were accommodated in the site's vast trailer park.

> ## "I think I can safely say that **nobody** understands **quantum mechanics.**"

Richard Feynman, 1964

form of the theory, which meant that its equations tended to generate impossible, or infinite, values.

Quantum probabilities

Something of a showman, Feynman had a talent for communicating complex ideas in intuitive ways, and he brought his particular skills to improving the QED theory. In 1948, he formulated a version of the theory that produced meaningful results, with the aid of what became known as Feynman diagrams. These useful graphics illustrated the concepts of electrons and photons moving in both time and space. Crucially, the particles in his diagrams could move not only forward but also backward in time.

Feynman's simple diagrams were in fact a mathematical model in which the probabilities of all possible events could be added up. Many of the probabilities added up to zero and cancelled each other out, leaving answers that made sense in real-world terms—at least, most of the time. The quantum universe is peculiar, and QED is, according to the title of one of Feynman's most popular books, a "strange theory of light and matter."

In Feynman diagrams, electrons are depicted by straight lines, while photon exchange between electrons is shown by wavy lines. Looking like doodles at first sight, the diagrams are in reality shorthand for extremely complicated equations. Today, particle physicists use them in a whole range of applications, including calculating possible outcomes of events in particle accelerators such as the Large Hadron Collider in Switzerland. As a result of Feynman's input, QED became one of the most successful theories in physics and was adopted as the model for theories of other basic forces, such as the nuclear force.

Nanotechnology and superfluidity

Feynman made his mark in other key areas of physics, too. He proposed a theory of "partons," a precursor to the modern concept of quarks (see box); helped to introduce nanotechnology; theorized about the future existence of quantum computing; and explained why, when cooled to absolute zero, helium flows without viscosity (internal friction). In 1986, Feynman formed part of a team investigating the *Challenger* disaster: the NASA space shuttle had exploded seconds after lift-off, killing all seven members of its crew. In a famous public demonstration, Feynman explained how the accident had happened using just a glass of icy water and a piece of rubber.

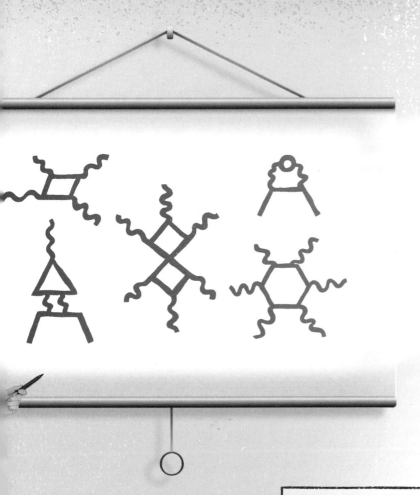

Feynman diagrams depict the most abstract concepts in physics but look like charming squiggles. Feynman even decorated his yellow RV with them. Their simplicity reflects Feynman's flair for explaining complex ideas—his series of physics lectures at the California Institute of Technology (Caltech) was made into a textbook that proved extremely popular.

MURRAY **GELL-MANN**

With wide-ranging talents, including languages, abstract thinking, and particle physics, Murray Gell-Mann was able to conceive the modern system for classifying subatomic particles.

In 1964, American polymath Gell-Mann (1929–) proposed a model for classifying the different types of subatomic particles. He distinguished two main groups of particles: fermions, the building blocks of matter; and bosons, which are force-carriers. To create his model, he had invented a hypothetical fundamental unit he called the "quark." Quarks, which are a type of fermion, group together to form hadrons (protons and neutrons). A few years later, the quark was isolated using a particle accelerator, proving Gell-Mann's model correct. He won the 1969 Nobel Prize in Physics, and his model became what is now known as the "standard model."

US oceanographer Henry Stommel combined his sharp intellect with a passion for the sea to transform his discipline into a major scientific field. He helped to explain the powerful forces that drive ocean currents, including the Gulf Stream.

Henry Stommel studied astronomy at Yale University, and taught there for 2 more years before joining the Woods Hole Oceanographic Institution (WHOI) in 1944. He later took up oceanography posts at Harvard and at the Massachusetts Institute of Technology (MIT), but returned to WHOI before the end of his career.

In 1948, Stommel published a groundbreaking paper showing how a combination of fluid dynamics, the Earth's curvature, and the Coriolis effect (in which moving objects are deflected to one side by the rotation of the Earth) causes stronger currents to be found on the western boundaries of ocean basins. His elegant solution explained the position and intensity of the Gulf Stream—a powerful, warm ocean current that runs up the western side of the Atlantic. The paper was characteristic of Stommel's intuitive grasp of complex phenomena, and it opened up a whole new area of study: the dynamics of large-scale ocean circulation.

Stommel and his colleagues developed his ideas into a wider theory of global ocean circulation with their work on thermohaline (temperature and salinity) effects. When sea ice forms at the poles, the seawater gets saltier and denser and sinks to the depths, pulling in warmer water behind it to create a conveyor belt of moving water looping through the world's oceans. Surface-level currents are predominantly wind-driven, but thermohaline forces drive deep-ocean circulation. Stommel later oversaw two of the most successful global studies of ocean processes.

The "Great Ocean Conveyor Belt," driven by variations in water temperature and salinity, is part of the global ocean circulation system envisioned by Stommel.

1920–1992

"**Science** is a voyage ... and an **expression** of the **human spirit.**"

Henry Stommel, 1989

HENRY
STOMMEL

Chinese-born theoretical physicist Yang Chen-Ning revolutionized the field of particle physics with his groundbreaking research into subatomic decay, for which he received the Nobel Prize in Physics.

After obtaining a PhD in nuclear physics from the University of Chicago, Yang Chen-Ning moved to Princeton to study subatomic decay (the transformation of one subatomic particle into others) in 1949. At the time, a law of physics called conservation of parity stated that it is not possible to distinguish between different directions in particle interactions. Yang's work led him to question this law.

In 1956, after extensive research with partner Tsung-Dao Lee, the pair proposed the law was invalid for certain so-called "weak" interactions, such as radioactive decay. They also suggested an experiment that might prove their proposition was correct. The experiment was successfully carried out by Chien-Shiung Wu (see p.253) later that year, and won Yang and Lee the Nobel Prize in Physics in 1957.

(see p.253)

"These **concepts** were **not dreamed up**. They were **natural and real**."

Shiing-Shen Chern, 1975

MILESTONES

US SCHOLARSHIP
Leaves China after being awarded a scholarship at the University of Chicago to study physics in 1946.

PARTICLE THEORY
Proposes the Yang-Mills theory on elementary particles with his colleague Robert Mills in 1954.

BREAKTHROUGH PAPER
Publishes paper in 1956 to prove conservation of parity is fallible; awarded Nobel Prize for his work in 1957.

RESEARCH DIRECTOR
Becomes director of the theoretical physics research center at Stony Brook University in 1965.

YANG CHEN-NING

BENOIT

MANDELBROT

Polish-born French mathematician Benoît Mandelbrot founded fractal geometry, which created a new way of collecting and analyzing data.

Benoît Mandelbrot emigrated to France in 1936, and gained a PhD in mathematics from the University of Paris. He took a role at IBM in New York in 1958 and devised the concept of fractal geometry, which detects an ordered logic in seemingly random or chaotic things. This form of geometry is based on symmetry found in repeating patterns and can be applied to natural phenomena, such as coastlines or tree bark, as well as analyses of data, such as noise patterns in economics and weather patterns. It created a new way of visualizing data that has since been applied in many fields—including medicine, engineering, and cosmology—and has diverse applications such as helping to understand financial markets, predict earthquake patterns, and diagnose disease.

The Mandelbrot set is a fractal plotted using a group of complex numbers. When rendered on computers, it reveals an infinitely repeating shape and pattern.

Canadian Ernest McCulloch was a hematologist who dreamed of finding a cure for leukemia. When a radiation experiment produced an odd result, he recognized it as something entirely new.

Ernest McCulloch was born and raised in Toronto. He studied medicine at Toronto University and qualified as a doctor in 1948. After a period of time at the Lister Institute, a medical research charity in London, UK, he returned to Toronto and worked as a research fellow in pathology at Sunnybrook Hospital. In 1957, he joined the newly created Ontario Cancer Institute, a research division of Princess Margaret Hospital. There he began the main body of his research into blood formation, both normal and malignant, and soon joined forces with James Till. Their work on stem cells would transform cancer research and give rise to regenerative medicine. During his career, McCulloch was a key figure in Canadian medical research and served on many national and international advisory committees.

MILESTONES

DEPARTMENTAL HEAD
Becomes the Head of Hematology at the newly formed Ontario Cancer Institute in 1957.

PIVOTAL PUBLICATION
Publishes findings of stem cell research with Till and Andrew Becker in the journal *Nature* in 1963.

LEUKEMIA STUDIES
In the 1970s, focuses his stem cell research on analyzing the mechanisms of human leukemia.

NATIONAL HERO
Along with Till, is inducted into the Canadian Science and Engineering Hall of Fame in 2010.

"McCulloch was the 'big picture' thinker."

Tak Wah Mak, 2018

ERNEST McCULLOCH

Physicist James Till turned his attention to cancer research and, while working on radiation treatment, he and Ernest McCulloch made the extraordinary discovery of stem cells.

James Till grew up in a farming community in the rural border town of Lloydminster, Canada. He won a scholarship to the University of Saskatchewan to study physics, and upon graduation went to Yale University, where he completed a PhD in biophysics in 1957. Till then took up a research fellowship at the University of Toronto and was recruited to the Ontario Cancer Institute. In 1958, he began his work with McCulloch.

Later in his career, Till's research interests expanded in relation to cancer treatment. His studies included research ethics and patient quality of life, as well as the role of the internet in patient advocacy and support.

McCulloch and Till discovered that most bones of the human body contain unspecialized cells called adult stem cells. Residing in the inner marrow, they are able to develop into any blood cell.

MILESTONES

RETURN TO CANADA
Declines an assistant professorship at Yale in 1957, in favor of a post at the University of Toronto.

SEMINAL PAPER
In 1961, with McCulloch, publishes a paper on the "radiation sensitivity of ... mouse bone marrow cells."

LEADING SCHOLAR
Becomes a Fellow of the Royal Society of Canada, the senior national council of scholars, in 1969.

PRESTIGIOUS AWARD
Presented with the Albert Lasker Award for Basic Medical Research in 2005, jointly with McCulloch.

JAMES
TILL

McCulloch and Till's combination of skills, in cellular biology and biophysics, proved powerful. In 1961, while carrying out research into bone marrow cells, the two discovered the first clue to the existence of a special type of cell that became known as the stem cell. These cells have a unique potential to become other, specialized types of cells and can also self-renew and multiply themselves.

Unexpected developments

The game-changing discovery came about quite by chance; McCulloch described it as a case study in "the importance of serendipity in scientific research." McCulloch was interested in the medical potential of nuclear radiation and wanted to explore its effects on cancer cells. His project involved irradiating mice, and McCulloch, a medical doctor, needed a physicist with the necessary expertise in radiation—Till.

The two began measuring the radiation sensitivity of the mice's bone marrow cells by exposing the mice to radiation that would be lethal if they did not receive a transplant, then replacing their bone marrow cells with donor cells. At the time, bone marrow transplantation was a new treatment known to replenish crucial blood cells. However, Till and McCulloch had also exposed the donor marrow in their transplants to varying doses of radiation. By monitoring the mice's response to the transplants, the scientists could measure how many of the donor cells had survived the differing radiation doses, and therefore the number of live marrow cells the mice had received.

The team repeated the experiment and this time carried out autopsies on the mice 10 days after the transplants.

> "Major breakthroughs in science involve unexpected findings. You **can't expect** the **unexpected.**"

James Till, 2018

One Sunday morning, McCulloch went into the laboratory to assess the tissue samples from the autopsies and was surprised to see a number of small bumps on the dead mice's spleens. The number of bumps, or nodules, was in exact proportion to the number of live marrow cells the mice had received.

Colonies of clones

Till and McCulloch needed to find out whether each of these nodules had grown from a single cell or from many cells. So, with Till's PhD student Andrew Becker, they again irradiated the donor cells with the aim of creating a marker, or trace. In a few of the irradiated donor cells, the radiation caused small changes to the chromosomes. These altered chromosomes acted as a marker, in that they would be present in any new cells that had grown from them. After isolating and examining hundreds of cells from a new set of spleen nodules, the team found that each nodule had grown from a single cell. In other words, they were clones. Till and McCulloch referred to the original cells as "colony-forming units."

More remarkably, the colonies of cells in each nodule were a mixture of those that develop into the three types of blood cells: red cells, white cells, and platelets. Till and McCulloch knew this was exciting, but they were cautious. In 1961, they published their findings in a plainly titled

HEART

LUNGS

BLOOD VESSELS

SKIN

SHINYA **YAMANAKA**

By "reprogramming" mature cells, Yamanaka was able to turn them back into stem cells.

In 2006, Yamanaka (1962–) introduced 24 specific genes into adult skin cells in mice and made them revert to an immature stem-cell state. He also showed that these cells were "pluripotent," meaning they can become any type of cell. Previously, it was thought that only embryonic stem cells were pluripotent. For his "induced pluripotent stem cells," or iPS cells, Yamanaka jointly received the Nobel Prize in Physiology or Medicine in 2012.

paper that referred to the "radiation sensitivity of normal mouse bone marrow cells." It was largely ignored.

In 1963, Till, McCulloch, and Becker published further results in the journal *Nature*. This time, they got attention. They went on to publish even more studies and clearly showed that bone marrow contains special cells that can reproduce and also develop into other types of cells, with specialized functions. We now know that, in adults, stem cells exist in many organs and tissues; this is how the body heals or regenerates damaged tissue. Their discovery was a vital advance in biological science and has huge potential for treating disease.

Stem cells exist in the body at all stages of a person's life. Embryonic stem cells can develop into any type of cell. Adult stem cells are more specialized— although recent research suggests they may be more flexible than once thought.

AFTER THE **MARROW** TRANSPLANTS, THE **MICE** WERE EXAMINED FOR **10 TO 11**

DAYS

EACH **MOUSE** SPLEEN NODULE CORRESPONDED TO **10,000** BONE MARROW **CELLS**

STEM CELL RESEARCH **ACCELERATED** THE DEVELOPMENT **OF BONE MARROW** TRANSPLANTS

TU
YOUYOU

Chinese pharmacologist Tu Youyou painstakingly researched thousands of compounds for their antimalarial properties at the end of the 1960s—and finally isolated artemisinin, which became the main malaria treatment worldwide. Her discovery eventually earned her a Nobel Prize in 2015.

MILESTONES

INSPIRING ILLNESS
Contracts tuberculosis as a teenager in 1937, and becomes inspired to research medicines.

EARLY RESEARCH
Studies schistosomiasis as part of her first job in 1955, and gains insights into parasite diseases.

HEADS TEAM
Aged 39, builds a team to find a new antimalarial in 1969, after thousands of compounds fail screening.

ANCIENT KNOWLEDGE
In 1971, reads a 1,600-year-old recipe that helps her find an effective way to extract artemisinin.

Tu Youyou was born in Ningbo on China's east coast. Despite an interruption to her education when she had tuberculosis (TB) for 2 years, she went on to study pharmacology at Peking University, where her tutor introduced her to herbal medicine. In 1955, she joined the newly formed Academy of Traditional Chinese Medicine. Here, she researched *Lobelia chinensis* (Chinese lobelia) as a remedy for schistosomiasis (bilharzia), an infection carried by a parasitic worm that was common in parts of China.

Malaria research
Caused by the single-cell *Plasmodium* parasite and transmitted by female *Anopheles* mosquitoes, malaria was on the rise throughout the world, including in China, despite the World Health Organization's global malaria eradication program (1955–1968). Death rates were rising and new treatments were urgently needed due to increasing resistance of the parasite to existing drugs, such as chloroquine. In 1969, as there had not been any new medicines discovered, the Chinese government appointed Tu as head of the secretive "Project 523" on Hainan Island to source new antimalarial drugs from traditional Chinese medicines. Within 3 months, Tu and her team had gathered more than 2,000

"No doubt, traditional Chinese medicine provides a **rich resource."**

Tu Youyou, 2015

Tu is the chief scientist at the China Academy of Traditional Chinese Medicine, Beijing. The institution currently serves as the World Health Organization Collaborating Centre for Traditional Medicine.

"Approach research through the integration of diversified disciplines."

Tu Youyou, 2015

herbal, mineral, and animal substances with medicinal potential. They then tested 380 extracts from 200 plants for their ability to kill malaria-causing *Plasmodium* parasites in mice. An extract of *Artemisia annua* (qinghao or sweet wormwood) initially seemed favorable, but the results were inconsistent.

New preparation technique
A standard way of preparing many plants in Chinese medicine is to make a decoction, in which the root is simmered

GE HONG

in water. Tu wondered if this was the correct method for qinghao—and looked in reference materials, including Ge Hong's *A Handbook for Prescriptions for Emergencies*, written more than 1,600 years ago. He wrote, "A handful of qinghao, immersed in two liters of water, wring out the juice and drink it all." This made Tu think that heat from boiling the herb in a decoction might be destroying the active ingredients and that the aerial parts (the leaves and stalks) rather than the roots might have active properties, as these aerial parts could be wrung out. She therefore redesigned the experiment and extracted the active ingredients from the leaves and stems using a low-temperature method involving water, ethanol, and ethyl ether in 1971.

Her method worked, and Tu went on to isolate the active compound artemisinin, which successfully treated malaria in mice. Artemisinin kills the malaria parasite by damaging its membranes and interfering with the process by which it digests its host's hemoglobin. When analyzed chemically, artemisinin proved to be a sesquiterpene lactone, a different compound to any other known antimalarial drug.

Further success

Tu's team then had to carry out clinical studies on humans. In July 1972, she and two colleagues offered to evaluate

Tasked with finding a new antimalarial, Tu visited physicians across China practicing traditional herbal medicine. She collected more than 2,000 recipes within 3 months, including Ge Hong's (right) prescription for qinghao, which was key to Tu's successful discovery.

SATOSHI **OMURA**

Research by Japanese biochemistry professor Omura led to the creation of the drug ivermectin.

Omura (1935–) was a microbiologist at Kitasato University. Like Tu, he sourced medicines from natural substances, including the bacterium *Streptomyces avermitilis*. Further exploration of its potent antimicrobial properties eventually produced ivermectin, a key antiparasite medicine for many diseases, including river blindness and elephantiasis. Omura won the Nobel Prize in Physiology or Medicine in 2015 for his research.

artemisinin's toxicity by taking it themselves. They then ran a successful clinical trial in 1973, and the drug was licensed for use in 1986.

While continuing her research into artemisinin, Tu developed a second compound, dihydroartemisinin, that was 10 times more potent. This meant a smaller dose of the new compound would achieve the same effect. It was licensed by the Chinese Ministry of Health in 1992. Fast-acting artemisinin-based drugs are now widely given as part of combination therapy for malaria.

In 2011, Tu won the prestigious Lasker Award and, in 2015, she shared the Nobel Prize in Physiology or Medicine with Satoshi Omura and William C. Campbell, who had also used natural sources to create antiparasitic medicines. She was the first Chinese woman ever to win a Nobel Prize and the first Chinese person to win the Nobel Prize in Medicine. Today, Tu continues to work as chief scientist at the China Academy of Traditional Chinese Medicine.

SCIENTISTS WORLDWIDE TESTED MORE THAN **240,000 COMPOUNDS** PRIOR TO THE DISCOVERY OF **ARTEMISININ**

RECEIVED THE **NOBEL PRIZE** WHEN SHE **WAS 85**

ARTEMISININ COMBINATION THERAPY TREATS **198 MILLION PEOPLE** ANNUALLY

British primatologist Jane Goodall is the world's primary expert on chimpanzees. Her study of their behavior, which lasted more than 55 years, led to discoveries that fueled her ongoing campaign for animal welfare and conservation.

Aged just 26, Jane Goodall traveled to Gombe National Park in Tanzania for a 6-month program to study the behavior of chimpanzees. The Kenyan paleontologist Louis Leakey secured funds for this trip, having seen Goodall's potential and already sent her to London in 1958 to study primate anatomy and behavior.

Kindred spirits

With natural patience, tenacity, and empathy, Goodall overcame initial rejection from the animals and was eventually accepted into the chimpanzee community. Her 6-month placement extended into decades of detailed research that revealed for the first time the personalities, emotions, and social behaviors demonstrated by chimpanzees. Her key discoveries included that chimpanzees are omnivorous, hunt cooperatively, and wage war with one another—traits previously believed to be displayed only by humans. Furthermore, she noted their close similarities to humans in terms of their emotional behavior (both positive and negative), intelligence, social structure, and relationship-building.

In 1977, Goodall founded the Jane Goodall Institute, now a global organization that promotes her conservation campaign, and today, she is a world leader for the protection of chimpanzees and their environments. As a pioneering woman in a male-dominated world, she is an inspirational figure for a new generation of scientists.

Goodall observed that chimpanzees utilize tools, such as stripping a stick to gather food, revealing similarities between chimpanzee and human behavior.

JANE GOODALL

1934–

VALENTINA TERESHKOVA

Valentina Tereshkova is a Soviet cosmonaut and engineer who in 1963 became the first woman and the first civilian to fly in space.

After leaving school at age 16, Valentina Tereshkova worked in a textile factory and took up parachute jumping. Following the early Soviet victory over the US in the "space race"—achieved by sending the first man into space—she volunteered for the Soviet space program in 1961. After 16 months of training, in which she was exposed to extreme isolation and zero-gravity conditions, Tereshkova became the first woman to fly in space. She was launched in the space capsule Vostok 6 on June 16, 1963, and completed 48 orbits of Earth during 71 hours of flying. She never flew in space again but was made a Hero of the Soviet Union in 1966 and gained a doctorate in engineering in 1977.

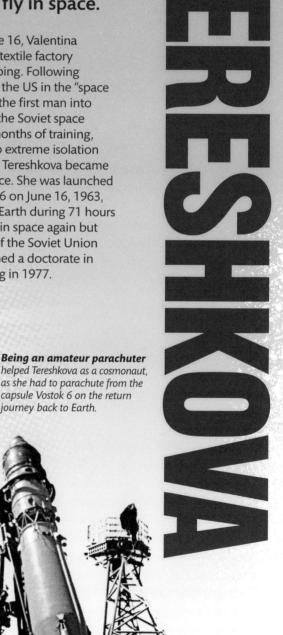

Being an amateur parachuter helped Tereshkova as a cosmonaut, as she had to parachute from the capsule Vostok 6 on the return journey back to Earth.

A pioneer for women in the field of ophthalmology, American Patricia Bath is an inventor, laser scientist, and ophthalmologist who has made huge global advances in the prevention of blindness.

Born in Harlem, New York, Patricia Bath studied medicine at Howard University and graduated in 1968. She was the first African American to complete a residency (postgraduate medical training) in ophthalmology at New York University, and went on to study corneal transplantation and keratoprosthesis (a procedure in which a cornea is replaced with an artificial one) at Columbia University. In 1974, she became associate professor of surgery and ophthalmology at UCLA and Charles R. Drew University. A fierce proponent of the idea that eyesight should be a basic human right, Bath co-founded the American Institute for the Prevention of Blindness in 1976 and launched "community ophthalmology," an outreach program bringing primary care to poor areas. In 1986, Bath invented the Laserphaco Probe, a laser that enabled more precise removal of cataracts. Bath lectured widely and continues to promote advanced ophthalmologic techniques, as well as provide vital, sight-saving treatments in developing nations.

MILESTONES

SETS PRECEDENT
Accepts position as first female ophthalmologist at UCLA's Jules Stein Eye Institute in 1975.

KEY INVENTION
Completes the Laserphaco Probe in 1986; receives a patent for her invention 2 years later.

TRAINING DIRECTOR
In 1983, is first US woman to set up and direct an ophthalmology residency training program.

"To restore sight is the ultimate reward."

Patricia Bath

PATRICIA
BATH

STEPHEN

HAWKING

The renowned theoretical physicist Stephen Hawking used mathematics to reveal the nature of "black holes" and to explore how the Universe began. He was also a gifted science popularizer who enabled millions around the world to deepen their understanding of space-time, gravity, and humanity's place in the cosmos.

Stephen William Hawking grew up in the town of St. Albans, north of London. In 1963, while at Cambridge University studying for a doctorate in astronomy, he was diagnosed with ALS, a form of incurable motor neurone disease. Although his mental faculties were unaffected, Hawking was confined to a wheelchair from 1969. In 1985, the disease robbed him of the use of his voice, and thereafter he communicated via a computer linked to a speech synthesizer, which he controlled using his facial muscles.

Mind over matter

The term "black hole" was coined in 1967, although a German physicist, Karl Schwarzschild, had hypothesized their existence in the 1910s. It was thought that as matter is dragged into a black hole, X-rays are created, and in 1964, astronomers were able to detect "X-ray binaries" close to star-forming regions in the Milky Way, indicating a possible black hole.

While at Cambridge, Hawking became fascinated with black holes, not least because he thought that they could yield insights into the very beginning of the Universe. He studied the work of mathematician Roger Penrose (see p.300), which proved that if Einstein's general theory of relativity is correct, the death and collapse of a giant star could culminate in a "singularity"— a point in space-time that is infinitely dense and infinitely small. In his PhD thesis, Hawking suggested that this process could run in reverse, and that

A BRIEF HISTORY OF TIME
FROM THE BIG BANG TO BLACK HOLES

STEPHEN W. HAWKING
INTRODUCTION BY CARL SAGAN

Hawking's first popular science book attempted to make theories about the past, present, and future of the Universe accessible to nonspecialists. It was a commercial and critical success, taking him and his work into the mainstream.

MILESTONES

WRITES PhD THESIS
In "Properties of expanding universes," written in 1966, he states that the Universe started from a singularity.

BLACK HOLES
Proves that gravitational singularities may exist while working with Roger Penrose in 1970.

KEY DISCOVERY
Draws on quantum theory in 1974 to predict that black holes emit heat and eventually evaporate.

RECEIVES HONOR
Becomes Cambridge's Lucasian Professor of Mathematics in 1979, a chair once held by Isaac Newton.

COMMUNICATES WORK
Publishes *A Brief History of Time* in 1988. It is an instant hit and launches him to cultural stardom.

the Universe could have started from a singularity. He compared the beginning of the Universe to the South Pole, saying: "To ask what happened before the beginning of the Universe would become a meaningless question, because there is nothing south of the South Pole."

Hawking radiation

Following a debate with Israeli physicist Jacob Bekenstein in 1972, Hawking refined his black hole theories to take into account the laws of quantum mechanics, which deals with the behavior of matter and energy on the scale of atoms and subatomic particles. These laws state that throughout the Universe, pairs of particles and antiparticles are constantly meeting and annihilating each other. Hawking calculated that when this occurs near the event horizon (the visible edge of a black hole), the negatively charged particle of the pair falls into the black hole and the positively charged half of the pair streams away from the black hole's edge as radiation—which became known as Hawking radiation. According to Hawking, the black hole loses energy and mass in a process called black hole evaporation as the negatively charged particles fall into it. Over eons, the black hole continues to leak radiation and particles until it disappears in a final explosion.

In 1974, Hawking published his calculation showing both how the Universe began and how black holes evaporate. By combining insights from the theory of the infinitely large (Einstein's general theory of relativity)

ROGER **PENROSE**

British mathematician and theoretical physicist Roger Penrose is known for his work on black holes and the Big Bang theory.

In the 1960s, Penrose (1931–) was a reader and then professor in applied mathematics at Birkbeck College, London, where he developed his work on black hole singularities. In 1969, he collaborated with Stephen Hawking to prove that black holes can arise from the gravitational collapse of massive stars, and their paper was published by the Royal Society of London in 1970. Penrose also developed twistor theory, which is a key tool in quantum theory, and created a way of mapping the regions of space-time around a black hole. Featuring so-called Penrose diagrams, this mapping technique makes it possible to visualize the effect that gravity has on an object approaching a black hole.

Hawking's discovery that black holes are not really black, but rather emit thermal radiation due to quantum effects, was one of the most important findings of 20th-century physics. Initially, his formula for Hawking radiation was rejected by many of his peers, but it has since become universally accepted.

> **"Black holes** are **stranger than** anything dreamed up by **science fiction** writers, **but** they are firmly matters of **science fact."**

Stephen Hawking, 2008

and the theory of the infinitely small (quantum mechanics), he produced a significant clue in modern physics' ongoing quest to find a single, unified "theory of everything" to explain the nature of the Universe.

Fame beckons
In 1979, Hawking was appointed Lucasian Professor of Mathematics at Cambridge University, a post he held until 2009. Undeterred by his disabilities, he continued to research black holes and received many accolades and high honors for this and other contributions to physics. He also wrote several popular science books, including *A Brief History of Time*, which was a runaway success and made him a household name. In his final years, he traveled widely, giving public lectures about his work to captivated audiences.

CALCULATED THAT **TIME, SPACE, ENERGY,** AND **MATTER** BEGAN FROM A **SINGLE DENSE POINT** OF **ENERGY—A SINGULARITY**

A BRIEF HISTORY OF TIME STAYED ON THE *SUNDAY TIMES* **BESTSELLER LIST** FOR **237 WEEKS**

WON THE BREAKTHROUGH PRIZE IN FUNDAMENTAL PHYSICS, **WORTH $3 MILLION**

JOCELYN BELL BURNELL

MILESTONES

PUBLISHES FINDINGS

Observes a pulsar in 1967. Antony Hewish reports her discovery in a *Nature* article the following year.

PHYSICS PRIZE

Although not named in the award herself, her supervisor receives the Nobel Prize in 1974.

OFFICIAL RECOGNITION

Made a Dame Commander of the British Empire (DBE) in 2007 for her services to astronomy.

As a young postgraduate student, Jocelyn Bell Burnell was the first scientist to detect pulsars, opening up a new branch of astrophysics.

Born in Belfast in Northern Ireland, Jocelyn Bell Burnell developed a flair for science at boarding school, where she realized "how easy physics was." In her teenage years, she resolved to work in astronomy.

After graduating from Glasgow University in 1965, Bell Burnell began a PhD thesis on twinkling quasars at Cambridge University. Under the supervision of Antony Hewish, a reader in astrophysics, she helped to design and build an array of radio telescopes that probed deep into space to detect quasars, a recently discovered form of high-energy cosmic radiation. In November 1967, her instruments detected unexpected signals: radio waves pulsating every 1.337 seconds, coming from a fixed point in space. Similar signals were detected in other parts of space over the following 2 months.

Bell Burnell and Hewish concluded that the radio waves came from the radiation beam of a rapidly rotating, strongly magnetized neutron star. Each rotation emitted a "pulse," giving the name "pulsar" to this entirely new class of star.

The suspiciously precise radio waves led Bell Burnell to playfully apply the working title LGM-1 (Little Green Men-1) until the term "pulsar" was coined in 1968.

Immunologist and molecular biologist Tak Wah Mak discovered how T-cells, which are part of the body's immune system, recognize antigens and advanced our molecular knowledge of cancer cells.

Tak Wah Mak was born in southern China in 1946, the son of a wealthy silk trader, and spent his childhood in Hong Kong. With his family, he then moved to the US, where he studied at the University of Wisconsin. He gained a PhD in biochemistry at the University of Alberta, Canada, in 1968.

Mak's research on T-cells, a type of white blood cell produced by the thymus gland, led to him solving a longstanding problem in immunology—how receptors on the surface of T-cells recognized antigens (substances that provoke the body's immune response). His breakthrough came in 1983 with the discovery that the genes of T-cell receptors used for antigen recognition have a unique genetic sequence and origin. This discovery opened up a new chapter in the study and treatment of immune-system diseases. Mak then applied molecular techniques to learn more about immune responses to cancer.

MILESTONES

BECOMES FELLOW
Accepts a fellowship at the Ontario Cancer Institute in 1972, following the completion of his doctorate.

T-CELL FINDINGS
Details the genetic encoding of human T-cell receptors in *Nature* journal's March 1984 edition.

GENETIC RESEARCH
Uses experiments with mice to isolate genetic function for cancer and immunology studies in 1988.

TARGETED TREATMENT
Identifies a chemical to block the fuel supply of Hodgkin's lymphoma, a rare form of cancer, in 1999.

"We can **never give up the fight** against [cancer]."

Tak Wah Mak, 2011

TAK WAH
MAK

TIM
BERNERS-LEE

British computer scientist Tim Berners-Lee invented the World Wide Web—now a multimedia library, workplace, and shopping and social space that can be accessed via the internet. He set it up in 1991 and gave it to the world for free.

MILESTONES

EARLY PROGRAM
Writes ENQUIRE in 1980, a computer program to connect CERN academics and their projects.

LANDMARK PAPER
Submits "Information Management: A Proposal" in 1989, outlining a linked hypertext data system.

WORLD WIDE WEB
Launches the World Wide Web at CERN in 1991, including HTML, HTTP, and the first browser.

RECEIVES KNIGHTHOOD
Knighted in 2004 by Queen Elizabeth II for services to the global development of the internet.

Born in London in 1955, Tim Berners-Lee was interested in computers from an early age. His parents met at UK company Ferranti, where they worked as computer programmers on the Mark I, the world's first general-purpose business computer. Berners-Lee went to Oxford University in 1973 to study physics, and while he was there, he built his first homemade computer using a second-hand television, some logic gates, and a processor.

After graduating, Berners-Lee worked as a software engineer for British companies before joining CERN—Europe's nuclear research lab, now best known as the home of the Large Hadron Collider—as a consultant in 1980. It was while grappling with the communication demands of the international laboratory's multilingual staff that he hit on the idea that would transform the way the world communicates.

Connecting the world

In 1980, Berners-Lee wrote ENQUIRE, a program that could store information and track the connections between people and projects at CERN. Although it only existed on Berners-Lee's computer, ENQUIRE demonstrated that the documents collected on the internet—a physical network of computers that had been growing steadily since its origins in 1963—could one day be accessed by anyone with a computer. Largely promoted by academics and the US Department of

Tim Berners-Lee's vision of "universal access to a large universe of documents" has been far exceeded, yet his dream of an open, democratic Web has been threatened by growing political and commercial interests online.

Defense, the internet had originally offered little or no ordinary public access. Berners-Lee left CERN in 1980, but returned in 1984 as a fellow. At that time, CERN was the largest internet node in Europe, and to modify it, Berners-Lee proposed the World Wide Web in 1989. This was a program that could enable users to access information via the internet using a "browser."

Open access

From the start, Berners-Lee urged CERN to make the World Wide Web available to the world for free by putting it into the public domain instead of patenting it—a very generous decision that cost him billions. Fittingly, his first step once the Web went live on April 30, 1991, was to publish a manifesto that included detailed instructions for programmers on how they could build their own websites. Effectively the first ever webpage, this document outlined

> # "The Web does not just connect machines, it connects people."

Tim Berners-Lee, 2008

two ingeniously simple standards called HTML (hypertext markup language) and HTTP (hypertext transfer protocol). HTML is a standard way of coding webpages so that any computer can display them; HTTP is a "language" that allows the web browsers on people's computers to communicate with the web servers that store websites in HTML form.

Berners-Lee's invention soon gathered pace, partly due to the simplicity of HTML and HTTP, but also thanks to the development of secure, easy-to-use web browsers. By the mid-1990s, millions of people—home users, universities, scientists, governments, and businesses— were interacting, creating, selling, and shopping with the World Wide Web. In 1994, Berners-Lee set up the World Wide Web

Thanks to Tim Berners-Lee's creation of the World Wide Web, people are able to communicate via email and to conduct business electronically across the globe.

Consortium (W3C), which continues to develop the Web's open standards. But from the earliest days of his invention, he has consistently advocated for a fair, democratic, and truly "World Wide" Web that brings political, social, and economic benefits to everyone on the planet.

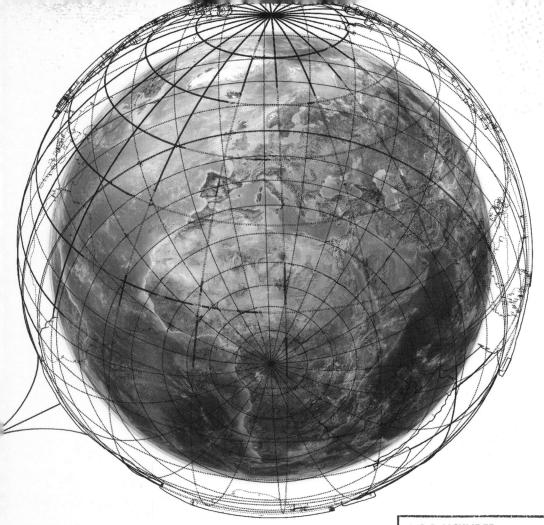

PUBLISHED THE **FIRST EVER WEBPAGE** ON **AUGUST 6, 1991**

THERE ARE NOW **MORE THAN 4 BILLION** INTERNET USERS

CHOSE NOT TO RECEIVE **ROYALTIES** FROM **THE WEB**

J. C. R. **LICKLIDER**

Born in Missouri, Licklider was an early champion for computer power.

Head of Information Processing at the US government's Advanced Research Projects Agency (ARPA) from 1962, Licklider (1915–1990) was an early force behind computer technology. He commissioned Arpanet—the Intergalactic Computer Network—in 1963, a project that connected research computers together. Procedures for transferring data were fixed by 1978 and became compulsory in 1983, effectively creating the internet.

"I MAY HAVE INVENTED THE WEB, BUT ALL OF YOU HAVE HELPED TO CREATE WHAT IT IS TODAY. NOW IT IS UP TO ALL OF US TO BUILD THE WEB WE WANT—FOR EVERYONE. "

Tim Berners-Lee
Three Challenges for the Web, 2017

◄ *Tim Berners-Lee* poses with the logo of the
World Wide Web Consortium (W3C) in 1995.

DIRECTORY

Science has become both increasingly diverse and increasingly connected. Researchers across the globe collaborate to increase human knowledge in every field and seek to bring their different disciplines closer together in pursuit of bigger, broader, more useful theories.

FRED HOYLE
1915–2001

British astrophysicist Fred Hoyle worked on the origin of elements in stars. He postulated that most chemical elements could be created step-by-step by nuclear reactions within large stars. Hoyle was also a proponent of the steady state theory. This states that as the Universe expands, its average density is kept constant as new matter is continuously created. Hoyle coined the phrase "Big Bang" for the main rival theory during a popular radio talk but used the term disparagingly. The steady state theory was largely discredited in the 1960s. In later life, Hoyle took particular interest in the presence of organic molecules in comets, which he believed had brought life to Earth.

IRENE UCHIDA
1917–2013

A Canadian medical geneticist, Irene Uchida instigated work on how genes and chromosomes affect health. She investigated chromosome differences in various genetic conditions and congenital abnormalities, including congenital heart disease and Down syndrome, in which people have an extra chromosome (47 rather than 46) in their cells. She also examined how X-rays affect chromosomes.

KATHERINE JOHNSON
1918–

Award-winning American mathematician Katherine Johnson played a key role in calculating the trajectory for many space missions, including those of Alan Shepard, the first American in space. She began work at the National Advisory Committee for Aeronautics (NACA), a forerunner of NASA, in 1953, manually carrying out calculations for the engineers there. She helped to plan the launch site and flight path for the Apollo 11 Moon mission in 1969, and later worked on the US space shuttle program. She retired from NASA in 1986.

YVONNE BRILL
1924–2013

Born near Winnipeg, Canada, Yvonne Brill started work with the Douglas Aircraft company in 1945. She went on to invent a propulsion system that keeps unmanned spacecraft in stationary orbit. She also worked on the thrusters for weather satellites, rocket designs used in US Moon missions, and the Mars observer. While working for NASA from 1981 to 1983, she worked on a rocket engine for the space shuttle.

ABDUS SALAM
1926–1996

Born in Pakistan, Abdus Salam studied mathematics and physics at the University of Cambridge, UK. He proposed the electroweak theory, which explains how the weak interaction between subatomic particles works and how it could be unified with the electromagnetic force. His theory postulated the existence of messenger particles for the weak interaction, with different messengers being exchanged in different instances of the interaction. His theories were proved correct at CERN (the European Organization for Nuclear Research) in 1973, and he was jointly awarded the Nobel Prize in Physics with Sheldon Lee Glashow and Steven Weinberg in 1979.

VERA RUBIN
1928–2016

During the 1970s, Vera Rubin discovered that stars orbiting in the outer part of a spiral galaxy travel as fast as those close to the center. This was contrary to Newton's laws of gravity and implied that each galaxy had a halo of invisible matter whose gravitational force affected the outer stars. In 1970, Rubin showed this to be true of the Andromeda Galaxy; by 1985, she had examined 60 spiral galaxies and realized it was a general phenomenon. Rubin's work convinced astronomers that dark matter exists. She received many prestigious awards, including the National Medal of Science in 1993.

PETER HIGGS
1929–

Born in Newcastle, England, Peter Higgs studied physics at King's College London. In 1960, he began work at the University of Edinburgh. Higgs is best known for his involvement in developing a theory that explains the origin of the mass of elementary particles (subatomic particles). According to this theory, these particles obtain their masses by interacting with a field, called the Higgs field, that permeates space. He predicted that this field should produce its own type of particle, which has since become known as the Higgs boson. The search for the Higgs boson spawned the world's largest science project, the Large Hadron Collider (LHC) at CERN, and scientists here went on to discover a particle with the predicted characteristics of the Higgs boson in 2012. A year later, Higgs received the Nobel Prize in Physics, sharing it with François Englert for the work they carried out in 1964.

ERIC KANDEL
1929–

Born in Vienna in 1929, Eric Kandel fled with his family to the US as anti-Semitism became rife. He was interested in the cellular basis of behavior and studied medicine at New York University, receiving his degree in 1956. In 2000, along with two others, he was awarded the Nobel Prize in Physiology or Medicine for work on the role synapses (nerve junctions) play in memory and learning. Through investigations on the sea slug (which has a simple nervous system with large cells that are easy to study), he helped to clarify the different processes behind short- and long-term memory and showed that they are equally applicable in humans.

MARIO J. MOLINA
1943–

Born in Mexico City, Mario Molina studied to become a chemical engineer. In 1974, he and colleague Sherwood Rowland showed that CFCs (chlorofluorocarbon compounds) released from aerosols, foams, and refrigerants destroy the ozone layer, which helps protect humans from the Sun's ultraviolet radiation. This work led directly to banning or limiting the use of CFCs and earned Molina and Rowland a share of the 1995 Nobel Prize in Chemistry.

LAP-CHEE TSUI
1950–

Canadian medical geneticist Lap-Chee Tsui was born in Shanghai and moved to Canada in 1977. He jointly found the fault in the gene responsible for cystic fibrosis. The life-threatening disease causes mucus produced by the body to be very thick—the mucus particularly affects the lungs. The identification of the faulty gene paved the way for new treatment options based on correcting the gene or replacing it.

MAE CAROL JEMISON
1956–

The first African American woman to travel in space, Mae Carol Jemison went into orbit in the space shuttle *Endeavour* in 1992. Jemison had graduated from medical school in 1981 before serving as a medical officer with the Peace Corps in West Africa. In 1985, she applied to the US space agency, and in 1987, she became one of the 15 candidates selected from more than 2,000 applicants to train as an astronaut. Jemison was the science mission specialist on the space shuttle and a coinvestigator of two bone cell research experiments performed there.

DONNA STRICKLAND
1959–

Canadian optical physicist Donna Strickland paved the way for the shortest and most intense laser pulses ever created. In 1985, along with Gérard Mourou from France, she developed a technique called chirped pulse amplification (CPA), and for this she and Mourou were jointly awarded the Nobel Prize in Physics in 2018. Strickland became associate professor in the Department of Physics and Astronomy at the University of Waterloo in 1997, where she now leads the ultrafast laser research team. CPA has several uses, including in laser eye surgery and in some cancer therapies.

CAROL GREIDER
1961–

Most of American molecular biologist Carol Greider's research has been on telomeres and the enzyme telomerase. Telomeres are the segments of DNA that occur at the ends of chromosomes and play an important role in cell life span. In 2009, along with molecular biologist Elizabeth Blackburn and geneticist Jack W. Szostak, Greider was awarded the Nobel Prize in Physiology or Medicine. She found that inhibiting the telomerase in cancer cells slows tumor growth, making it a potential anticancer drug development.

E

F

G

H

ACKNOWLEDGMENTS

Dorling Kindersley would like to thank the following: Gillian Andrews for additional design assistance, Chauney Dunford for additional editorial assistance, Jacqueline Street-Elkayam for additional pre-production assistance, and:

Indexer: Helen Peters
Proofreader: Steph Lewis
Jackets Editorial Coordinator: Priyanka Sharma
Managing Jackets Editor: Saloni Singh

PICTURE CREDITS

The publisher would like to thank the following for their kind permission to reproduce their photographs:

(Key: a-above; b-below/bottom; c-center; f-far; l-left; r-right; t-top)

1 Alamy Stock Photo: Science History Images (c). **2 Alamy Stock Photo:** Pictorial Press Ltd. **8 Alamy Stock Photo:** colaimages. **10 Alamy Stock Photo:** Pictorial Press Ltd. **11 Alamy Stock Photo:** INTERFOTO (b). **12 Dreamstime.com:** Fortton (bc). **12-13 Dreamstime.com:** David Carillet (bc). **13 123RF.com:** dimitriosp (c/Statues); Viktoriya Sukhanova (b). **Dreamstime.com:** Ruslan Gilmanshin (c). **Wellcome Collection http:// creativecommons.org/licenses/by/4.0/:** (tc). **14 Dreamstime.com:** Nickolayv (br). **15 Alamy Stock Photo:** Peter Horree. **16 123RF.com:** Sergey Buzuevskiy / Aquarius83men (br); Panagiotis Karapanagiotis (br/Statue). **Alamy Stock Photo:** The Picture Art Collection (crb/Tremoctopus). **Dreamstime.com:** Markus Gann / Magann (cr/Sun); Skallapendra (tc); Torian Dixon / Mrincredible (cr); Yury Shirokov (crb). **17 123RF.com:** Kichigin Aleksandr (cr); darkbird (c, cb); milkos (b). **Alamy Stock Photo:** age fotostock (bc). **Dorling Kindersley:** Steve Teague (clb). **Dreamstime.com:** Evgeny Drobzhev (Background). **18 Alamy Stock Photo:** colaimages. **20 123RF.com:** kubais (br). **21 123RF.com:** Ilya Akinshin (b); wektorygrafika (br). **Alamy Stock Photo:** Interfoto (tc); World History Archive (c). **Dreamstime.com:** Vtorous (br). **22 Dreamstime.com:** Awcnz62 (b). **Getty Images:** Science & Society Picture Library (br). **23 Alamy Stock Photo:** Historic Collection. **24 Alamy Stock Photo:** Science History Images. **25 Alamy Stock Photo:** Science History Images (b). **26 123RF.com:** Sebastian Kaulitzki (cr). **Dreamstime.com:** Ljubisa Sujica (cl); Vadymvdrobot (br). **Wellcome Collection http://creativecommons.org/licenses/by/4.0/:** (cra). **26-27 Dreamstime.com:** Norbert Buchholz (cb). **27 Dreamstime.com:** Anolis01 (cl); Igor Zakharevich (bl). **Getty**

Images: Jumpstart Studios (clb). **Wellcome Collection http://creativecommons.org/licenses/by/4.0/:** (cr). **28 Getty Images:** DeAgostini. **30-31 Alamy Stock Photo:** Classic Collection 3 (tc). **30 Dorling Kindersley:** Whipple Museum of History of Science, Cambridge (br). **31 Alamy Stock Photo:** History and Art Collection. **32 Dreamstime.com:** Anna Lurye (crb). **33 123RF.com:** dbelashova (clb); nikolae (cra); kchung (cl); Ljubisa Sujica (cl/Head); skarintut (c). **Alamy Stock Photo:** The Picture Art Collection (bc). **Dreamstime.com:** Denis Barbulat (cra/Icons). **Library of Congress, Washington, D.C.:** (tc). **34 123RF.com:** Pitchayarat Chootai / pitchayarat2514 (br/ Book). **Dreamstime.com:** Nikolai Grigoriev (bl). **Getty Images:** DEA / A. CASTIGLIONI / De Agostini (cl). **35 Getty Images:** Chris Bradley. **36 123RF.com:** Alexandr Blinov (tr). **37 akg-images:** Heritage Images (bc). **Dreamstime.com:** Zatletic (bl). **40 Dreamstime.com:** Georgios Kollidas / Georgios. **42-43 Fotolia:** HP Photo (tc). **42 Dreamstime.com:** Georgios Kollidas. **44 Alamy Stock Photo:** Granger Historical Picture Archive (cb). **Wellcome Collection http://creativecommons.org/licenses/by/4.0/:** (cr). **44-45 Dreamstime.com:** Yanta (b). **45 Wellcome Collection http://creativecommons.org/licenses/by/4.0/:** (cl, cb). **46-47 123RF.com:** Illia Balla / noirion (bc). **46 akg-images:** Bilwissedition (br). **47 123RF.com:** Lukasz Janyst. **48 Wellcome Collection http://creativecommons.org/licenses/by/4.0/:** (crb). **49 123RF.com:** nikolae (br). **Dreamstime.com:** Torian Dixon / Mrincredible. **50 Dreamstime.com:** Georgios Kollidas / Georgios. **51 Wellcome Collection http://creativecommons.org/licenses/by/4.0/:** (bl). **52-53 Dreamstime.com:** Clearviewstock (b). **52 123RF.com:** Krzysztof Kruz (cr). **iStockphoto.com:** ZU_09 (bl). **53 Alamy Stock Photo:** Chronicle (tr). **Dreamstime.com:** Mexrix (b); Patrimonio Designs Limited / Patrimonio (bc); Rainledy (crb). **55 Alamy Stock Photo:** National Geographic Image Collection. **56 Wellcome Collection http://creativecommons.org/licenses/by/4.0/. **57 Getty Images:** Oxford Science Archive / Print Collector / Hulton Archive (bl). **58 Wellcome Collection http://creativecommons.org/licenses/by/4.0/:** (tc). **59 Dreamstime.com:** Markus Gann / Magann (cb); Torian Dixon / Mrincredible; Natthawut Punyosaeng / Aopsan (t). **60 Wellcome Collection http://creativecommons.org/licenses/by/4.0/. **61 Wellcome Collection http://creativecommons.org/licenses/by/4.0/:** (bl). **64 Wellcome Collection http://creativecommons.org/licenses/by/4.0/. **66 Bridgeman Images:** Private Collection / © Look and Learn. **67 Alamy Stock Photo:** Science History Images (bl). **68 Library of**

Congress, Washington, D.C.: Tissandier Collection (cb). **69 Alamy Stock Photo:** World History Archive (tc). **Dorling Kindersley:** The National Railway Museum, York / Science Museum Group (c). **Dreamstime.com:** Andrii Zhezhera (ca). **70 Alamy Stock Photo:** Science History Images (r). **71 Alamy Stock Photo:** Colport. **Dorling Kindersley:** COPPOLA STUDIOS (tl). **72-73 Dorling Kindersley:** The Science Museum, London (bc). **73 Alamy Stock Photo:** Pictorial Press Ltd (tc). **Dreamstime.com:** Nerthuz (ca); Topvectorstock (cra). **Wellcome Collection http://creativecommons.org/licenses/by/4.0/:** (crb). **74 Alamy Stock Photo:** Photo 12. **76 Dreamstime.com:** Georgios Kollidas / Georgios. **77 Alamy Stock Photo:** World History Archive (bl). **78 Wellcome Collection http://creativecommons.org/licenses/by/4.0/:** (crb). **79 123RF.com:** Hong Li (br); Georgios Kollidas (bl). **Dreamstime.com:** Torian Dixon / Mrincredible (cb). **81 Alamy Stock Photo:** Chronicle. **82 Alamy Stock Photo:** Ian Dagnall (tl). **82-83 akg-images:** Erich Lessing (bc). **84 Alamy Stock Photo:** Classic Image (ca); Ian Dagnall (bl). **Dreamstime.com:** Dave Bredeson (crb/Key); Anna Velichkovsky (crb). **Getty Images:** Apic / Retired (tr). **85 123RF.com:** Kittisak Taramas (crb). **Dreamstime.com:** Ulkass (t); Anna Velichkovsky (tc). **86 123RF.com:** Tamara Kulikova (br). **Wellcome Collection http://creativecommons.org/licenses/by/4.0/:** (br/Plants). **87 Wellcome Collection http://creativecommons.org/licenses/by/4.0/. **88 Dreamstime.com:** Amador Garcia Sarduy / Loverpower (crb). **Wellcome Collection http://creativecommons.org/licenses/by/4.0/:** (cb). **89 123RF.com:** nikolae (cr); Andrejs Pidjass / NejroN (ca); nrey (clb); rasslava (cl). **Dreamstime.com:** Nadezhda Khvatova (cr/leaves). **Wellcome Collection http://creativecommons.org/licenses/by/4.0/:** (cb). **90 Alamy Stock Photo:** The Natural History Museum. **92 Alamy Stock Photo:** The Print Collector. **93 Alamy Stock Photo:** The Natural History Museum (bl). **94 Library of Congress, Washington, D.C.:** (br). **95 iStockphoto.com:** denisk0. **96 Dreamstime.com:** AWesleyFloyd (crb). **97 Dreamstime.com:** Novi Elysa (cra); Jktu21 (Background). **Wellcome Collection http://creativecommons.org/licenses/by/4.0/:** (l, crb). **98 Alamy Stock Photo:** Science History Images. **99 Dreamstime.com:** Luisa Vallon Fumi (br). **100 Wellcome Collection http://creativecommons.org/licenses/by/4.0/. **101 Alamy Stock Photo:** INTERFOTO (bl). **102-103 Dreamstime.com:** Taras Kulyk (b). **103 123RF.com:** lightvisionftb (cl). **Dreamstime.com:** Soleilc (crb). **Wellcome Collection http://creativecommons.org/licenses/by/4.0/:** (tr). **104 Wellcome Collection http://creativecommons.org/licenses/by/4.0/. **104-105 Alamy Stock**

Photo: Granger Historical Picture Archive (bc). **106 Wellcome Collection http://creativecommons.org/licenses/by/4.0/:** (br). **107 Alamy Stock Photo:** The Print Collector. **108 123RF.com:** Gerard Koudenburg (br); Sviatlana Zykava (cr). **Alamy Stock Photo:** Science History Images (cl/ Edward). **Dreamstime.com:** Mohdhaka (cl). **Science Photo Library:** Dr. Klaus Boller (cr/ TEM). **109 Wellcome Collection http://creativecommons.org/licenses/by/4.0/:** (tr). **110 123RF.com:** Georgios Kollidas. **111 Science Photo Library:** Emilio Segre Visual Archives / American Institute Of Physics (cb). **112 123RF.com:** bakai (clb). **Wellcome Collection http://creativecommons.org/licenses/by/4.0/:** (tc). **113 Alamy Stock Photo:** The Picture Art Collection (l/Symbols on Balloons, br). **Wellcome Collection http://creativecommons.org/licenses/by/4.0/:** (r). **114 Wellcome Collection http://creativecommons.org/licenses/by/4.0/. **115 Getty Images:** Boyer / Roger Viollet (br). **116 123RF.com:** Marguerite Voisey (bc). **117 Alamy Stock Photo:** Dotted Zebra (bl, br). **Library of Congress, Washington, D.C.:** James W. Black, (tr). **120 Dreamstime.com:** Georgios Kollidas. **122 Dreamstime.com:** Georgios Kollidas. **123 Wellcome Collection http://creativecommons.org/licenses/by/4.0/:** (bl). **124 Library of Congress, Washington, D.C.:** Brady-Handy Collection (tc). **124-125 Dreamstime.com:** Bozhdb (c). **125 123RF.com:** aleksanderdn (cb). **Dreamstime.com:** Michael Beer (cb/Hair dryer); Andres Rodriguez / Andresr (cb); Pixelgnome (clb); Dannyphoto80 (tc); Kevynbj (cb/Desk Fan); Pioneer111 (cb/oven); Valentyna Chukhlyebova. **126 Wellcome Collection http://creativecommons.org/licenses/by/4.0/. **127 Alamy Stock Photo:** Granger Historical Picture Archive (crb). **Getty Images:** Science & Society Picture Library (b). **128 Dorling Kindersley:** The Science Museum (bl). **Dreamstime.com:** Ratz Attila (c). **Wellcome Collection http://creativecommons.org/licenses/by/4.0/:** (ca). **128-129 Dreamstime.com:** Alexander Limbach (t). **129 Dreamstime.com:** Davooda (cr); Razvan Ionut Dragomirescu / Twindesigner (tc). **Wellcome Collection http://creativecommons.org/licenses/by/4.0/:** (cb). **130 Dorling Kindersley:** The Science Museum (bl). **131 Alamy Stock Photo:** Balfore Archive Images (cl). **132 Alamy Stock Photo:** Pictorial Press Ltd. **133 Alamy Stock Photo:** INTERFOTO (bl). **134 Wellcome Collection http://creativecommons.org/licenses/by/4.0/:** Barraud (br). **135 123RF.com:** Brandon Alms / macropixel (ca); Andrey Pavlov (c); Csak Istvan. **Dreamstime.com:** Linnette Engler (ca/Yellow-faced Grassquit). **Getty Images:** Science & Society Picture Library (br, clb, c/Small Tree Finches, cl/Ground

Finch, ca/Warbler Finches). **Wellcome Collection http://creativecommons.org/licenses/by/4.0/:** Sims (tr). **136 akg-images**. **138 Alamy Stock Photo:** The Print Collector. **139 Alamy Stock Photo:** North Wind Picture Archives (bc). **141 Getty Images:** Oxford Science Archive / Print Collector (Background). **Wellcome Collection http://creativecommons.org/licenses/by/4.0/**. **142 Alamy Stock Photo:** The History Collection (cb); Profimedia.CZ a.s. (br). **142-143 123RF.com:** Andrey Bodrov (cb). **143 Alamy Stock Photo:** gameover. **Getty Images:** Sheila Terry / Science Photo Library (c). **144 Wellcome Collection http://creativecommons.org/licenses/by/4.0/:** (br). **145 Alamy Stock Photo:** Pictorial Press Ltd. **146 Wellcome Collection http://creativecommons.org/licenses/by/4.0/:** (cb). **147 Alamy Stock Photo:** Science History Images (cla); World History Archive (cl). **Dorling Kindersley:** Science Museum, London (bc). **Dreamstime.com:** Max776 (Background); Leigh Prather (c); Peterguess (cb). **Wellcome Collection http://creativecommons.org/licenses/by/4.0/:** (clb). **148 Getty Images:** Science & Society Picture Library (br). **149 Dreamstime.com:** Nicku. **150-151 Science Photo Library:** Carlos Clarivan. **150 Alamy Stock Photo:** colaimages (cb). **152 Alamy Stock Photo:** The History Collection. **153 Alamy Stock Photo:** INTERFOTO (br). **154 123RF.com:** Mr_Vector (bl). **154-155 123RF.com:** Patrick Marcel Pelz. **155 Alamy Stock Photo:** Science History Images (tr). **Dreamstime.com:** Dmitriy Melnikov / Dgm007 (cr). **157 Alamy Stock Photo:** Science History Images. **158 Dreamstime.com:** Georgios Kollidas. **159 Alamy Stock Photo:** Historical Images Archive (b). **162 Alamy Stock Photo:** IanDagnall Computing. **164-165 Alamy Stock Photo:** Science History Images (bc). **165 Wellcome Collection http://creativecommons.org/licenses/by/4.0/**. **166 123RF.com:** nikolae (cla). **Dreamstime.com:** Sabuhi Novruzov (cl); Vvcephei (cb); Ksenia Palimski (bc); Texelart (br). **Wellcome Collection http://creativecommons.org/licenses/by/4.0/:** (crb). **167 Alamy Stock Photo:** Pictorial Press Ltd (cl). **Dreamstime.com:** Andrey Kiselev (bl). **Wellcome Collection http://creativecommons.org/licenses/by/4.0/:** (cr). **168 Science Photo Library:** Emilio Segre Visual Archives / American Institute Of Physics (br). **169 Alamy Stock Photo:** Science History Images. **170 123RF.com:** Andriy Popov (crb). **Alamy Stock Photo:** Science History Images (cb). **Getty Images:** Augenstein / ullstein bild (cr). **170-171 123RF.com:** Nadezhda Alkimovich (c/Light); Liubou Yasiukovich (c). **171 123RF.com:** Dashadima (bl). **173 Getty Images:** Ullstein bild. **174-175 Science Photo Library:** American Philosophical Society (bc). **175 123RF.com:** Polesnoy (Background). **Getty Images:** Heritage Images / Hulton Archive. **176 Alamy Stock Photo:** World History Archive. **177 Getty Images:** Universal History Archive (bl). **178 Science Photo Library:** American Philosophical Society (br). **179 Alamy Stock Photo:** Pictorial Press Ltd. **180 Alamy Stock Photo:** Science History Images (cb). **181 123RF.com:** 3ddock (b); Alekss (cl, c). **Dreamstime.com:** Rauf Aliyev (crb); Studiophotomh (cra). **182 Alamy Stock Photo:** IanDagnall Computing. **183 Wellcome Collection http://creativecommons.org/licenses/by/4.0/:** (bl). **184 Dreamstime.com:** Max776 (b). **Library of Congress, Washington, D.C.:** (tc). **185 123RF.com:** Belchonock (clb); iconmama (crb); Stepan Popov (b). **Alamy Stock Photo:** IanDagnall Computing (c); World History Archive (cl). **Dorling Kindersley:** Mark Cavaanagh (b/Radiation). **Dreamstime.com:** Bidouze St⊠phane (cb). **186 Alamy Stock Photo:** Science History Images (br). **187 Alamy Stock Photo:** GL Archive. **188 Dreamstime.com:** Destina156 (c). **Getty Images:** De Agostini / G. Cigolini (tr). **189 Alamy Stock Photo:** Archive Pics (tc). **190 Science Photo Library:** Prof. Peter Fowler. **192 Alamy Stock Photo:** Interfoto. **193 Getty Images:** Duane Howell / The Denver Post (bl). **194 akg-images:** (br). **195 Getty Images:** ullstein bild. **196 Alamy Stock Photo:** Science History Images (cb). **Dreamstime.com:** Fotyma (br). **Getty Images:** Bettmann (cl). **197 Alamy Stock Photo:** Christine Whitehead (c). **Dorling Kindersley:** Lego / Lucasfilm (cb/Explosion). **Dreamstime.com:** Andreus (cb/fission); Daseaford (bl); Geografika (cb). **198 Alamy Stock Photo:** Sputnik. **199 Getty Images:** Keystone-France / Gamma-Keystone (bc). **200 123RF.com:** Hans Christiansson (cb). **Alamy Stock Photo:** Science History Images (tc). **Getty Images:** Keystone-France / Gamma-Keystone (c). **202 Alamy Stock Photo:** Granger Historical Picture Archive. **204 Alamy Stock Photo:** Science History Images. **205 akg-images:** (bl). **206 akg-images:** Philip Hood / De Agostini Picture Lib. (tc/Cynognathus, ca). **Alamy Stock Photo:** Universal Images Group North America LLC / DeAgostini (c). **206-207 Colorado Plateau Geosystems Inc:** Paleogeography Globe Derived from an original map produced by Colorado Plateau Geosystems Inc. **207 Alamy Stock Photo:** Granger Historical Picture Archive (crb). **Dorling Kindersley:** Natural History Museum, London (ca, cla, ca/skull). **210 Alamy Stock Photo:** Granger Historical Picture Archive. **212 Alamy Stock Photo:** Nordicphotos (bc). **212-213 Alamy Stock Photo:** Keystone Press. **214-215 Dreamstime.com:** Kao Studio (c). **215 123RF.com:** Costasz (b). **Alamy Stock Photo:** The Print Collector (tc). **Dorling Kindersley:** Ahmad Firdaus Ismail (c). **Dreamstime.com:** Renaud Philippe (cr). **216 Alamy Stock Photo:** Flhc 61. **217 Science Photo Library:** Emilio Segre Visual Archives / American Institute Of Physics (bl). **218 Alamy Stock Photo:** Keystone Pictures Usa / Zumapress.com (br). **219 Getty Images:** Bettmann. **220 Alamy Stock Photo:** Rbm Vintage Images. **221 Alamy Stock Photo:** ClassicStock (bc). **222 Alamy Stock Photo:** Granger Historical Picture Archive (cb). **Dorling Kindersley:** Andrey Alyukhin (tl). **Dreamstime.com:** Richard Thomas (tc, tr). **222-223 Science Photo Library:** Take 27 Ltd. **223 Dreamstime.com:** Richard Thomas (tl). **224 Wellcome Collection http://creativecommons.org/licenses/by/4.0/**. **225 Alamy Stock Photo:** Walter Oleksy. **226 Science Photo Library:** Hagley Museum And Archive. **227 Alamy Stock Photo:** Heritage Image Partnership Ltd (bl). **228 Getty Images:** Francis Miller / The Life Picture Collection. **229 Science Photo Library:** American Philosophical Society (bl); Science Source (br). **230 Alamy Stock Photo:** INTERFOTO. **231 Getty Images:** Nina Leen / The Life Picture Collection (b). **232 Getty Images:** Corbis (bc). **232-233 Alamy Stock Photo:** Science History Images (bc). **233 Alamy Stock Photo:** Everett Collection Inc. **234-235 Dorling Kindersley:** Bethany Dawn (b/background). **234 123RF.com:** 3drenderings (bc); Sabri Deniz Kizil / jackrust (cl, crb/crane); Onur kocamaz (clb, cb/Construction Worker , crb/construction worker); Ranko Bojanovic (bl); Dmitry Kalinovsky (crb). **Dorling Kindersley:** Francesca Yorke / Bradbury Science Museum, Los Alamos (cb/Fat Man). **Dreamstime.com:** Dmitry Kalinovsky / Kadmy (cb). **235 Getty Images:** Fritz Goro / The Life Picture Collection (bl); Keystone-France / Gamma-Keystone (tr). **236-237 Getty Images:** Bettmann (bc). **237 Getty Images:** Bettmann. **238 Alamy Stock Photo:** Science History Images. **239 Alamy Stock Photo:** John Dambik (bl). **240 Alamy Stock Photo:** Science History Images (cb). **241 Alamy Stock Photo:** Science History Images (tr). **242 Alamy Stock Photo:** Science History Images (br). **243 Dorling Kindersley:** (Background). **Getty Images:** Library Of Congress / Interim Archives. **244 Getty Images:** The Asahi Shimbun. **245 Getty Images:** The Asahi Shimbun (bl). **246 Science Photo Library:** Emilio Segre Visual Archives (br). **247 Getty Images:** Keystone / Hulton Archive. **248 123RF.com:** samum (tr). **Dreamstime.com:** Geografika (bc). **248-249 123RF.com:** sergpet (b). **249 123RF.com:** iconmama (crb). **Dorling Kindersley:** Lego / Lucasfilm (bl). **Dreamstime.com:** Taiga (t/background). **Getty Images:** Universal History Archive / Uig (tc). **250 Alamy Stock Photo:** Science Photo Library (br). **Dreamstime.com:** Chris Dorney (bl). **251 Getty Images:** Bettmann. **252 123RF.com:** mopic (crb). **Alamy Stock Photo:** Keystone Press (bl). **253 Getty Images:** Gjon Mili / The Life Picture Collection. **254-255 Dreamstime.com:** Lance Bellers (bc). **255 Alamy Stock Photo:** Granger Historical Picture Archive; Ken Hawkins (Background). **256 Alamy Stock Photo:** Ken Hawkins (Background). **Dorling Kindersley:** Imperial War Museum, London (clb). **Getty Images:** Sspl (br). **257 Alamy Stock Photo:** Granger Historical Picture Archive (cr). **258 Getty Images:** Photo 12 / Uig (tc). **262 Alamy Stock Photo:** Pictorial Press Ltd. **264 Getty Images:** Bettmann (bl). **264-265 Getty Images:** Sspl (bc/Background). **265 Getty Images:** Bettmann (br). **267 Getty Images:** Salk Institute For Biological Studies (tc). **269 Science Photo Library:** A. Barrington Brown, © Gonville & Caius College. **270 Alamy Stock Photo:** Pictorial Press Ltd. **270-271 Bridgeman Images:** Private Collection (bc). **272 Alamy Stock Photo:** Science History Images. **273 123RF.com:** Sviatlana Sheina (br). **Dorling Kindersley:** Yulia Drozdova / incomible (cra). **Getty Images:** George Silk / The Life Picture Collection (cb). **274 Alamy Stock Photo:** Martin Shields (b). **275 Alamy Stock Photo:** World History Archive. **276 123RF.com:** iconmama (cl). **Dreamstime.com:** Teresa Kenney (bl). **277 Dreamstime.com:** Teresa Kenney. **Getty Images:** Ferdaus Shamim / Sygma (tr). **278 Getty Images:** Photo 12 / Uig. **279 Getty Images:** Corbis (b). **280 Dreamstime.com:** Fuzzbones (r). **Getty Images:** Joe Munroe / Hulton Archive (cra). **281 Alamy Stock Photo:** Keystone Press (crb). **Dreamstime.com:** Mastaka (cla). **283 Woods Hole Oceanographic Institute**. **284 Alamy Stock Photo:** Keystone Press. **285 Getty Images:** Laguna Design (cb); Roger Ressmeyer / Corbis / Vcg (cl). **286 Rex by Shutterstock:** Peter Foley / EPA / Shutterstock. **287 Rex by Shutterstock:** Peter Foley / EPA / Shutterstock (br). **Science Photo Library:** Dennis Kunkel Microscopy (clb). **288-289 Dreamstime.com:** Wektorygrafika (b). **289 123RF.com:** eladora (clb); Milosh Kojadinovich (l). **Getty Images:** Ramin Talaie / Corbis (tc). **Science Photo Library:** Dennis Kunkel Microscopy (clb/SEM). **290 Getty Images:** VCG. **291 Getty Images:** STR / AFP (bc). **292 123RF.com:** charmboyz (Background); Wavebreak Media Ltd (bc); photo5963 (bl). **Alamy Stock Photo:** Xinhua (cl). **Wellcome Collection http://creativecommons.org/licenses/by/4.0/:** (cr). **293 Getty Images:** The Asahi Shimbun (tc). **294 Dreamstime.com:** Norma Cornes (bc). **Getty Images:** Dlillc / Corbis / Vcg (br). **295 Alamy Stock Photo:** Francis Specker. **296 Alamy Stock Photo:** Itar-Tass News Agency (br). **Dreamstime.com:** Fmua (bl). **297 Getty Images:** Jemal Countess. **298 Alamy Stock Photo:** Danita Delimont. **298-299 123RF.com:** Illia Balla / noirion (bc). **299 Alamy Stock Photo:** sjbooks (bl). **300 Alamy Stock Photo:** Science History Images (cb). **300-301 Dreamstime.com:** Torian Dixon / Mrincredible (tc); Peter Jurik / Pitris; Vitalyedush (c); Natthawut Punyosaeng / Aopsan (background). **301 Dreamstime.com:** Elena Stebakova (crb). **302 Alamy Stock Photo:** Stocktrek Images, Inc (br). **Getty Images:** Colin McPherson / Corbis (bl). **303 Getty Images:** Jonathan Wong / South China Morning Post. **304 Alamy Stock Photo:** Wenn Rights Ltd. **305 Science Photo Library:** Cern (br). **306 Alamy Stock Photo:** Wenn Rights Ltd (cb/head). **Dreamstime.com:** Photoeuphoria / Jaimie Duplass (cb). **307 123RF.com:** Kirill Cherezov (t). **Getty Images:** Philip Preston / The Boston Globe (crb). **308 Science Photo Library:** Sam Ogden

All other images © Dorling Kindersley
For further information see: www.dkimages.com